Hesperia Supplement 29

THE ATHENIAN GRAIN-TAX LAW OF 374/3 B.C.

Hesperia Supplements

I*	Sterling Dow, *Prytaneis: A Study of the Inscriptions Honoring the Athenian Councillors* (1937)
II*	Rodney S. Young, *Late Geometric Graves and a Seventh Century Well in the Agora* (1939)
III*	Gorham Phillips Stevens, *The Setting of the Periclean Parthenon* (1940)
IV*	Homer A. Thompson, *The Tholos of Athens and Its Predecessors* (1940)
V*	William Bell Dinsmoor, *Observations on the Hephaisteion* (1941)
VI*	James H. Oliver, *The Sacred Gerusia* (1941)
VII*	Gladys R. Davidson and Dorothy Burr Thompson, *Small Objects from the Pnyx: I* (1943)
VIII*	*Commemorative Studies in Honor of Theodore Leslie Shear* (1949)
IX*	John V. A. Fine, *Horoi: Studies in Mortgage, Real Security, and Land Tenure in Ancient Athens* (1951)
X*	Lucy Talcott, Barbara Philippaki, G. Roger Edwards, and Virginia R. Grace, *Small Objects from the Pnyx: II* (1956)
XI*	James R. McCredie, *Fortified Military Camps in Attica* (1966)
XII*	Daniel J. Geagan, *The Athenian Constitution after Sulla* (1967)
XIII	James H. Oliver, *Marcus Aurelius: Aspects of Civic and Cultural Policy in the East* (1970)
XIV	John S. Traill, *The Political Organization of Attica* (1975)
XV	Stephen V. Tracy, *The Lettering of an Athenian Mason* (1975)
XVI	Merle K. Langdon, *A Sanctuary of Zeus on Mount Hymettos* (1976)
XVII	T. Leslie Shear Jr., *Kallias of Sphettos and the Revolt of Athens in 268 B.C.* (1978)
XVIII*	Livingston Vance Watrous, *Lasithi: A History of Settlement on a Highland Plain in Crete* (1982)
XIX	*Studies in Attic Epigraphy, History, and Topography Presented to Eugene Vanderpool* (1982)
XX	*Studies in Athenian Architecture, Sculpture, and Topography Presented to Homer A. Thompson* (1982)
XXI	John E. Coleman, *Excavations at Pylos in Elis* (1986)
XXII	Elizabeth J. Walters, *Attic Grave Reliefs That Represent Women in the Dress of Isis* (1988)
XXIII	Clairève Grandjouan, *Hellenistic Relief Molds from the Athenian Agora* (1989)
XXIV	Jeffrey S. Soles, *The Prepalatial Cemeteries at Mochlos and Gournia and the House Tombs of Bronze Age Crete* (1992)
XXV	Susan I. Rotroff and John H. Oakley, *Debris from a Public Dining Place in the Athenian Agora* (1992)
XXVI	Ira S. Mark, *The Sanctuary of Athena Nike in Athens: Architectural Stages and Chronology* (1993)
XXVII	*Proceedings of the International Conference on Greek Architectural Terracottas of the Classical and Hellenistic Periods, December 12–15, 1991*, edited by Nancy A. Winter (1994)
XXVIII	D. A. Amyx and Patricia Lawrence, *Studies in Archaic Corinthian Vase Painting* (1996)

* *Out of print*

Hesperia: Supplement 29

THE ATHENIAN GRAIN-TAX LAW OF 374/3 B.C.

BY

RONALD S. STROUD

THE AMERICAN SCHOOL OF CLASSICAL STUDIES AT ATHENS
PRINCETON, NEW JERSEY
1998

Out-of-print *Hesperia* supplements may be obtained as reprints from:
Swets & Zeitlinger
Backsets Department
P.O. Box 810
2160 SZ Lisse
The Netherlands

Library of Congress Cataloging-in-Publication Data
Stroud, Ronald S.
The Athenian grain-tax law of 374/3 B.C. / by Ronald S. Stroud
p. cm. — (Hesperia Supplement ; 29)
Includes bibliographical references and index.
ISBN 0-87661-529-9 (pbk. : alk. paper)
1. Taxation (Greek law). 2. Tariff on farm produce—Greece—Athens—History—To 1500. 3. Grain—Taxation—Greece—Athens—History—To 1500. 4. Greece—Economic conditions—To 146 B.C. I. Title. II. Series: Hesperia (Princeton, N.J.). Supplement ; 29.
KL4380.S77 1998
343.38′505585641331—dc21 98-34192
CIP

TYPOGRAPHY BY THE AMERICAN SCHOOL OF CLASSICAL STUDIES AT ATHENS PUBLICATIONS STAFF
6–8 CHARLTON STREET, PRINCETON, NEW JERSEY
PRINTED IN THE UNITED STATES OF AMERICA
BY EDWARDS BROTHERS, INC., ANN ARBOR, MICHIGAN

This work is dedicated, with love and deep gratitude,

to my mother

Florence Greenwood Stroud

CONTENTS

PREFACE

I HERE PRESENT THE EDITIO PRINCEPS of a long and well-preserved Athenian law of 374/3 B.C. inscribed on a marble stele that was found in the Agora Excavations in 1986. In addition to the full text, translation, and notes on readings, I have tried to provide a fairly complete commentary on the many parts of this document that contribute significant new information on the history, law, economy, topography, and public finance of Athens in the Classical period. I have also noted topics on which this inscription seems to me to open up new avenues for future research, some of them potential subjects for Ph.D. dissertations and beyond the scope of the present monograph.

T. Leslie Shear, Jr., then Director of the Agora Excavations, first announced the discovery of the inscription that forms the subject of this study in the *Newsletter of the American School of Classical Studies at Athens*, Spring 1987, page 8. This was duly noted in *SEG* XXXVI 146, and there have been brief allusions to the document in a few subsequent publications, but its complete text appears in the present work for the first time. My original plan to publish the law as an article in *Hesperia* proved unrealistic as I became increasingly aware of the bulk and complexity of the new material it contains. Encouraged by the Publications Committee of the American School of Classical Studies at Athens, I decided that the importance of this new find deserved the more expansive presentation of a separate monograph.

In the preface to *Corinth* VIII, iii: *The Inscriptions, 1926–1950*, Princeton 1966, page vi, J. H. Kent observed, "I have tried not to bother too many people in my research." Over the past few years I have followed exactly the opposite approach and have drawn upon the expert knowledge of many friends and colleagues by distributing numerous copies of the text of this inscription, by lecturing on the grain-tax law in a wide range of settings, and by sending a preliminary draft of parts or all of the present work to several scholars for their criticisms. In the end, however, final responsibility rests with the editor of a newly discovered inscription. It is one of the great joys of epigraphic research to prepare the editio princeps of an important inscription. In publishing the now-famous Athenian decree honoring Kallias of Sphettos, T. Leslie Shear, Jr., aptly expressed "the sense of rare privilege" felt by the epigraphist entrusted with an assignment of this sort.[1] But I have learned from experience that such an assignment also carries a heavy burden, for few such first editions are free from serious error. "The world of Attic epigraphy," my teacher W. Kendrick Pritchett once remarked, "is not a perfect one." It is inhabited by many who are quick to point out mistakes. That is how we move forward. No one can aspire to producing a definitive editio princeps of a complex and lengthy Greek inscription.[2] The impossibility of having the last word, however, cannot diminish the pleasure of having the first one. It is a real privilege to introduce this remarkable new inscription into the world of scholarship.

I am deeply indebted to my friend T. Leslie Shear, Jr., for giving up his own plans to work on this text and assigning it to me for publication. He and his successor as Director of the Agora Excavations, John McK. Camp II, have provided much encouragement and ideal conditions for study of the stone in the Stoa of Attalos. I have had the benefit of a transcription of the text by Shear and have gained much from discussions of readings directly from the stone with both these astute scholars.

[1] Shear 1978, p. vi.

[2] "Eine editio princeps kann kein opus perfectum sein": Engelmann and Knibbe 1989, p. viii.

Among others who have aided my research with their many helpful suggestions, I wish to mention Jean Bingen, Michael H. Jameson, Angelos P. Matthaiou, Harry W. Pleket, Malcolm B. Wallace, and the members of the Claremont Book Club of Berkeley, California.

Special gratitude is owed to the following for reading part or all of an earlier draft of this work and allowing me to profit from their perceptive criticisms: the late Sara B. Aleshire, Alan L. Boegehold, John McK. Camp II, Kevin Clinton, Edward E. Cohen, Peter D. A. Garnsey, Philippe Gauthier, Christian Habicht, Mogens H. Hansen, Edward M. Harris, Sally Humphreys, Léopold Migeotte, Benjamin Millis, W. Kendrick Pritchett, Mary B. Richardson, Adele Scafuro, and T. Leslie Shear, Jr.

I am grateful to Anne Hooton for preparing Figure 7.

I am indebted to Helen Conrad Stroud for assistance with the Indexes.

At the request of the editor, I have provided translations of all extended quotations of ancient Greek. Although they have greatly increased the bulk of this work, we hope that these translations will make it accessible to as many readers as possible. All translations are my own.

RONALD S. STROUD
Athens, June 1998

ILLUSTRATIONS

FIGURES

ABBREVIATIONS OF PERIODICALS AND SERIES PUBLICATIONS

AA = *Archäologischer Anzeiger*
AAA = Ἀρχαιολογικὰ ἀνάλεκτα ἐξ Ἀθηνῶν
AfPap = *Archiv für Papyrusforschung und verwandte Gebiete*
AHB = *Ancient History Bulletin*
AJA = *American Journal of Archaeology*
AJAH = *American Journal of Ancient History*
AJP = *American Journal of Philology*
AM = *Mitteilungen des Deutschen Archäologischen Instituts, Athenische Abteilung*
AncW = *The Ancient World*
AnnPisa = *Annali della Scuola normale superiore di Pisa, Classe di lettere e filosofia*
AntCl = *L'Antiquité classique*
Ἀρχ Ἐφ = Ἀρχαιολογικὴ Ἐφημερίς
ATL = *The Athenian Tribute Lists*, B. D. Meritt, H. T. Wade Gery, M. F. McGregor, eds., 4 vols., Cambridge, Mass./Princeton, N.J., 1939–1953
BAAH = Βιβλιοθήκη τῆς ἐν Ἀθήναις Ἀρχαιολογικῆς Ἑταιρείας
BCH = *Bulletin de correspondance hellénique*
BE = *"Bulletin Épigraphique"* in *REG*
BEFAR = *Bibliothèque des Écoles françaises d'Athènes et de Rome*
BICS = *Bulletin of the Institute of Classical Studies of the University of London*
BSA = *Annual of the British School at Athens*
CAH = *The Cambridge Ancient History*, J. B. Bury, S. A. Cook, F. E. Adcock, M. P. Charlesworth, and N. H. Baynes, eds., Cambridge 1928–1939
CAH^2 = *The Cambridge Ancient History*, J. Boardman, I. E. S. Edwards, N. G. L. Hammond, et al., eds., 2nd edition, Cambridge 1982–
CJ = *Classical Journal*
ClAnt = *Classical Antiquity*
ClMed = *Classica et mediaevalia*
CP = *Classical Philology*
CQ = *Classical Quarterly*
CRAI = *Comptes rendus des séances de l'Académie des inscriptions et belles-lettres*
Δελτ = Ἀρχαιολογικὸν Δελτίον
DictAnt = *Dictionnaire des antiquités grecques et romaines*, C. Daremberg and E. Saglio, eds., Paris 1877–1913
EMC = *Échos du monde classique*
EpAnt = *Epigraphica Anatolica*
FGrH = *Die Fragmente der griechischen Historiker*, F. Jacoby, ed., Berlin 1923–
GRBS = *Greek, Roman, and Byzantine Studies*
HSCP = *Harvard Studies in Classical Philology*
I. Délos = *Inscriptions de Délos*, Paris 1926–1937
IG = *Inscriptiones Graecae*
I. Iasos = *Die Inschriften von Iasos*, W. Blümel, ed., Cologne
JHS = *Journal of Hellenic Studies*
LIMC = *Lexicon Iconographicum Mythologiae Classicae*
Meiggs-Lewis, *GHI* = *A Selection of Greek Historical Inscriptions to the End of the Fifth Century B.C.*, R. Meiggs and D. M. Lewis, eds., Oxford
MemLinc = *Memorie: Atti della Accademia nazionale dei Lincei, Classe di scienze morali, storiche e filologiche*
MusHelv = *Museum Helveticum*
OGIS = *Orientis Graeci Inscriptiones Selectae*, W. Dittenberger, ed., Leipzig

PA = *Prosopographia Attica*, J. Kirchner, ed., Berlin
PCG = *Poetae Comici Graeci*, R. Kassel and C. Austin, eds., Berlin
PCPS = *Proceedings of the Cambridge Philological Society*
Πρακτικά = Πρακτικὰ τῆς ἐν Ἀθήναις Ἀρχαιολογικῆς Ἑταιρείας
ProcAfrClSoc = *Proceedings of the African Classical Society*
PTeb = *Tebtunis Papyri*, B. P. Grenfell et al., eds., I–IV, 1902–1976
QuadUrb = *Quaderni Urbinati di Cultura Classica*
RE = *Paulys Real-Encyclopädie der klassischen Altertumswissenschaft*, A. F. Pauly, G. Wissowa, W. Kroll, et al., eds., Stuttgart
REA = *Revue des études anciennes*
REG = *Revue des études grecques*
RHDFE = *Revue historique de droit français et étranger*
RhM = *Rheinisches Museum für Philologie*
RPhil = *Revue de philologie, de littérature et d'histoire anciennes*
RSA = *Rivista storica dell'antichità*
RSC = *Rivista di Studi Classici*
SBWien = *Sitzungsberichte, Österreichische Akademie der Wissenschaften in Wien, Philosophisch-historische Klasse*
SEG = *Supplementum Epigraphicum Graecum*, Leiden, Amsterdam
Syll.^{3} = *Sylloge Inscriptionum Graecarum*, W. Dittenberger, 3rd edition, Leipzig
SymOslo = *Symbolae Osloenses*
TAPA = *Transactions of the American Philological Association*
YCS = *Yale Classical Studies*
ZPE = *Zeitschrift für Papyrologie und Epigraphik*

THE ATHENIAN GRAIN-TAX LAW OF 374/3 B.C.

CHAPTER I

THE GRAIN-TAX LAW

THE AGORA EXCAVATIONS continue to contribute dramatically to our understanding of legislative procedure and the economy in 4th-century B.C. Athens. Following the discovery of Nikophon's now-famous law on silver coinage of 375/4 B.C.,[1] the American excavators have found another product of *nomothesia* preserved virtually intact on a beautiful marble stele. Remarkably, the Athenians enacted this law in 374/3 B.C., the year following Nikophon's legislation. This time the *nomothetai* turned their attention to the grain that Athens imported each year from the islands of Lemnos, Imbros, and Skyros. They authorized intricate provisions regulating a tax on this grain, which I will argue was a tax in kind. They ordered the wheat and barley produced by this tax to be transported at a specific time to the Peiraieus, and from there it was to be brought up to the city, stored in the temenos of Aiakos, and sold in the Agora by public officials newly appointed for this purpose. The law allocates the proceeds from the sale of this grain to the military fund. Rich in valuable new information about the Athenian grain trade, public finance, and political institutions, this inscription also provides ample evidence about tax-farming in Classical Athens. It brings back to public life a prominent politician whose career seemed to have ended several years before. It helps to solve a long-standing problem in the topography of the Athenian Agora. Some of its provisions, however, remain obscure and difficult to interpret. This inscription will clearly be the object of intensive scrutiny for many years to come, especially among students of the Athenian economy. My first edition cannot do full justice to the many facets of Athenian history the new text illuminates. My aim is to present a trustworthy text and a commentary that will help stimulate further research on what will become a classic document.

I begin with a physical description of the monument, which John Camp found on July 21, 1986, built into a repair of the east wall of the Great Drain where it passes the northeast corner of the Stoa Basileios. Its findspot is thus only a few meters north of the place where the stele bearing Nikophon's law on silver coinage was recovered from the fabric of the Great Drain. The new law is inscribed on a complete stele of fine-crystalled, white marble mended from two closely fitting fragments. The smoothly polished inscribed surface is discolored a deep brown, especially at the right, where there is also considerable water damage. Fortunately, the latter has removed only a few letters from the ends of some lines. It has also made the surface of the right side of the stele friable, thereby obscuring the original tooling. Clearly, however, the right side was squared off for the full thickness of the stele. On the roughly dressed left side, where the stele is very well preserved, there are clear traces of a claw chisel, but this surface was only squared off and dressed to a maximum depth of ca. 0.07 m. behind the inscribed face, and not for the full thickness of the stone. The stele is broken at the bottom, but not much of the original is likely to be lost. Smoothly polished surface continues for ca. 0.04–0.05 m. below the final line of text, but then we find traces of a claw chisel identical to those on the left side. The top edge of the rougher surface thereby created forms a horizontal line that probably marks the point where the stele was set into its base. The back, which is very roughly dressed, was clearly never meant to be seen; it is likely that the stele stood against a wall or some other structure. See Figure 1.

[1] Stroud 1974; *SEG* XXVI 72. The extensive bibliography on this document is accessible by working back from *SEG* XLII 88. See also Dreher 1995, pp. 90–106.

Above line 2, which contains the archon-dating formula, there is an oval molding, 0.01–0.014 m. in height, surmounted by a fascia, 0.015 m. high, on which line 1 is inscribed. Both moldings are carried around onto the left and right sides of the stele; on the left for only ca. 0.038 m., on the right for 0.105 m., almost the full thickness of the stone. The difference between the right and the left sides of the stele in both the treatment of the moldings and the squaring off of the surface suggests that while the right side may have been exposed to view, there was less concern about the left. Perhaps when the mason worked on the stele, he knew that it would be placed next to something on the left in such a way that the left side would not be fully visible.

Above the moldings on the front there is a slightly recessed panel roughly oblong in shape. Its surface is as smoothly polished as the inscribed portion of the stele, but I have been unable to detect any trace of letters, incised lines, or markings of any kind here. If, as seems likely, it was not simply left blank, the most plausible suggestion is that this panel once carried a painting. The only clue as to its subject is the irregularly contoured top of the stele above this panel. Although the photograph may give the impression that this top is broken, there is no doubt that it was intentionally dressed in antiquity somewhat crudely to form four, or perhaps more, roughly rounded projections. These are unequal in dimensions and not symmetrically arranged, except that the central one, the tallest, lies on the vertical axis of the stele. The significance of these projections is not immediately clear. Perhaps they represent the tops of some kind of irregular floral decoration. Alternatively, the panel may have held a painting that had something to do with the subject of the law inscribed below. It is not out of the question that the irregular contours at the top of the panel belong to heaps or sacks of grain.[2]

[2] I am grateful to Carol L. Lawton, Dina Peppas-Delmouzou, and Olga Palagia for helpful discussion about the top of the stele. For bags of money represented in relief at the top of Kleonymos' decree of 426/5 B.C. on the collection of tribute, *IG* I^3 68, see Meritt 1967, pl. II; Meyer 1989, pp. 248–249, no. A 3; Lawton 1995, p. 81, no. 1. For a list of painted panels at the top of stelai bearing Attic decrees, see Walbank 1987, pp. 267–268; Lawton 1995, p. 13, note 35. This topic of painted scenes at the top of inscribed stelai might repay further study.

FIG. 1. Agora I 7557

TEXT

Height, 1.105 m.; width, at molding, 0.45 m., below molding (level of line 2), 0.422 m., at level of line 61, 0.437 m., at bottom, 0.44 m.; thickness, 0.115 m.
Height of letters, lines 1, 3–61, 0.007 m., line 2, 0.01 m.
Agora inv. no. I 7557.

374/3 B.C.

⟨Θ⟩ ε ο ί
Ἐπὶ Σωκρατίδο ἄρχοντος
Νόμος περὶ τῆς δωδεκάτης τοῦ σίτου *vv*
τῶν νήσων *vacat*
Ἀγύρριος εἶπεν· ὅπως ἂν τῶι δήμωι σῖ[το]- ΣΤΟΙΧ. 31
ς ἦι ἐν τῶι κοινῶι, τὴν δωδεκάτην πωλ[εῖ]-
ν τὴν ἐν Λήμνωι καὶ Ἴμβρωι καὶ Σκύρω[ι κ]-
αὶ τὴν πεντηκοστὴν σίτο· ἡ δὲ μερὶς ἐκ̣[ά]-
στη ἔσται πεντακόσιοι μέδιμνοι, πυ[ρῶ]-
ν μὲν ἑκατόν, κριθῶν δὲ τετρακόσιοι· [κο]-
μιεῖ τὸν σῖτον κινδύνωι τῶι ἑαυτõ ὁ π̣[ρ]-
ιάμενος εἰς τὸν Πειραιᾶ καὶ ἀνακομι[ε]-
ῖ εἰς τὸ ἄστυ τὸν σῖτον τέλεσιν τοῖς α̣[ὐ]-
τõ καὶ κατανήσει τὸν σῖτον εἰς τὸ Αἰά̣[κ]-
ειον· στέγον δὲ καὶ τεθυρωμένον παρέ̣[ξ]-
ει τὸ Αἰάκειον ἡ πόλις καὶ ἀποστήσε[ι τ]-
ὸν σῖτον τῆι πόληι τριάκοντα ἡμερῶ̣ν [ὁ]
πριάμενος, ἐπειδὰν ἀνακομίσηι εἰς [ἄσ]-
τυ τέλεσι τοῖς αὑτõ· ἐπειδὰν δὲ ἀνακ[ομ]-
ίσει εἰς τὸ ἄστυ, ἐνοίκιον οὐ πράξει [ἡ π]-
όλις τοὺς πριαμένους· τοὺς πυροὺς ἀ̣[πο]-
στήσει ὁ πριάμενος ἕλκοντας πέντε ἑ̣[κ]-
τέ⟨α⟩ς τὸ τάλαντον, τὰς δὲ κρι⟨θ⟩ὰς ἑλκο[ύσ]-
⟨α⟩ς τὸν μέδιμνον τάλαντον ξηρὰς ἀπο̣σ[τ]-
ήσει καθαρὰς αἰρῶν, τὸ σ⟨ή⟩κωμα ἐπὶ ΤΗ[. .]-
ΩΝ<Η>Ι σηκώσας, καθάπερ οἱ ἄλλοι ἔμ[π]ορ[ο]-
ι· προκαταβολὴν οὐ θήσει ὁ πριάμε[ν]ο[ς ἀ]-
λλ' ἐπώνια καὶ κηρύκεια κατὰ τὴν [μ]ερ[ίδ]-
α εἴκοσι δραχμ⟨ά⟩ς· ἐγγυητ⟨ὰ⟩ς καταστήσ[ε]-
ι ὁ πριάμενος δύο κατὰ τὴμ μερίδα ἀξι[ό]-
χρεως, οὓς ἂν ἡ βουλὴ δοκιμάσηι· συμ[μορ]-
ία ἔσται ἡ μερὶς τρισχίλιοι μέδιμ̣[νοι],
ἓξ ἄνδρες· ἡ πόλις πράξει τὴν συμμορ[ία]-
ν τὸν σῖτον κ⟨α⟩ὶ παρ' ἑνὸς καὶ παρ' ἁπάν[τω]-
ν τῶν ἐν τῆι συμμορίαι ὄντων, ἕως ἂ̣ν τ[ὰ α]-

ὑτῆς ἀπολάβηι· αἱρείσθω δὲ ὁ δῆμος δ[έκ]-
⟨α⟩ ⟨ἄ⟩νδρας ἐξ Ἀθηναίων ἁπάντων ἐν τῆι [ἐκ]-
κλησίαι, ὅτανπερ τοὺς στρατηγοὺς α[ἱρ]-
ῶνται, οἵτινες ἐπιμελήσονται τοῦ σί[τ]-
ο· οὗτοι δὲ ἀποστησάμενοι τὸν σῖτοṿ κ[α]-
τὰ τὰ γεγραμμένα πωλόντων ἐν τῆι ἀγ[ορ]-
ᾶι, ὅταν τῶι δήμωι δοκῆι· πωλε͂ν δὲ μὴ ἐ[ξε]-
ῖναι ἐπιψηφίσαι πρότερον τοῦ Ἀνθεσ[τ]-
ηριῶνος μηνός· ὁ δὲ δῆμος ταξάτω τὴν τ[ι]-
μὴν τῶν πυρῶν καὶ τῶν κριθῶν ὁπόσου χ[ρ]-
ὴ πωλε͂ν τοὺς αἱρεθέντας· τὸν δὲ σ⟨ῖ⟩τον [ο]-
ἱ πριάμενοι τὴν δωδεκάτην κομισάντω-
ν πρὸ τοῦ Μαιμακτηριῶνος μηνός· οἱ δὲ α̣-
ἱρεθέντες ὑπὸ τοῦ δήμου ἐπιμελούσθω-
ν, ὅπως ἂν κομίζηται ὁ σῖτος ἐν τῶι χρόṿ-
ωι τῶι εἰρημένωι· ἐπειδὰν δὲ ἀποδῶντα-
ι οἱ αἱρεθέντες τὸν σῖτον, λογισάσθω[ν]
ἐν τῶι δήμωι καὶ τὰ χρήματα ἡκόντων φ[έ]-
ροντες εἰς τὸν δῆμον καὶ ἔστω στρατι[ω]-
τικ⟨ὰ⟩ τὰ ἐκ τõ σίτο γενόμενα· τὴν δὲ προ[κ]-
αταβολὴν τὴν ἐκ τῶν νήσων μερίσαι το[ὺ]-
ς ἀποδέκτας καὶ τῆς πεντηκοστῆς, ὅσο[ν]-
περ πέρυσιν ⟨η⟩ὗρεν ἐκ τοῖν δυοῖν δεκάτ[.]-
ιν, τὸ μὲν νῦν εἶναι εἰς τὴν διοίκησι[ν κ]-
αὶ τὸ λοιπὸν μὴ ⟨ἀ⟩φαιρεῖν τὼ δύο δεκάτ[.]
ἐκ τῶν κατ⟨α⟩βαλλομένων χρημάτων *vvvv*

Fig. 2. Agora I 7557, lines 1–28

Fig. 3. Agora I 7557, lines 23–52

Fig. 4. Agora I 7557, lines 31–61

TRANSLATION

Gods.

In the archonship of Sokratides.

Law concerning the $8\frac{1}{3}$% tax (*dodekate*) on the grain of the islands.

(**Line 5**) Agyrrhios moved: In order that there may be grain for the people (*demos*) in the public domain, sell the $8\frac{1}{3}$% tax (*dodekate*) (which originates) in Lemnos, Imbros, and Skyros, and the 2% tax (*pentekoste*) in terms of grain. (**8**) Each portion will consist of 500 measures (*medimnoi*), 100 of wheat and 400 of barley. (**10**) The buyer of the tax (*priamenos*) will convey the grain to the Peiraieus at his own risk and he will transport the grain up to the city at his own expense and he will heap up the grain in the Aiakeion. (**15**) The state (*polis*) will make available the Aiakeion in watertight condition and provided with a door. The buyer of the tax (*priamenos*) is to weigh out the grain for the state (*polis*) within thirty days, whenever he has conveyed it up to the city at his own expense. (**19**) Whenever he has conveyed it up to the city, the state (*polis*) will not exact rent from the buyers of the tax (*priamenoi*). (**21**) The buyer of the tax (*priamenos*) will weigh out the wheat at a weight of a talent for five sixths (*hekteis*) and the barley at a weight of a talent per measure (*medimnos*), dry and free from darnel, filling up the measuring table (*sekoma*) to the [rim?], just as the other merchants (*emporoi*). (**27**) The buyer of the tax (*priamenos*) will not make a down payment but (he will pay) sales taxes (*eponia*) and herald's fees (*kerykeia*) of twenty drachmai per portion. (**29**) The buyer of the tax (*priamenos*) will provide per portion two solvent guarantors, whomever the Council (*Boule*) has approved. (**31**) The portion held by a group of six men (*symmoria*) will consist of 3,000 measures (*medimnoi*). (**33**) The state (*polis*) will exact the grain from the group (*symmoria*), both from one man and from all who are in the group, until it recovers what belongs to it.

(**36**) Let the Assembly (*demos*) elect ten men from all the Athenians in the meeting (*ekklesia*) at which they elect the generals (*strategoi*), who are to take care of the grain. (**40**) After these men have seen to the weighing out of the grain in accordance with the written instructions, let them sell it in the marketplace (*agora*) at whatever time seems best to the Assembly (*demos*), but it is not permitted to put the sale of the grain to a vote before the month of Anthesterion. (**44**) Let the Assembly (*demos*) set the price at which the elected men must sell the wheat and barley. Let the buyers (*priamenoi*) of the $8\frac{1}{3}$% tax (*dodekate*) transport the grain (to the Aiakeion) before the month of Maimakterion. (**48**) Let the men elected by the Assembly (*demos*) make sure that the grain is transported in the specified time. (**51**) When the men who have been elected have sold the grain, let them render their accounts in the Assembly (*demos*) and let them come into the Assembly (*demos*) carrying the money, and let the amount that is realized from the (sale of the) grain be assigned to the military fund (*stratiotika*). (**55**) Let the Receivers (*apodektai*) allocate the down payment from the islands and of the 2% tax (*pentekoste*) exactly as much as was fetched last year from the two tenths (*dekatai/oi/a*). (**59**) For the present it is to belong to the financial administration (*dioikesis*), and in the future let them not take the two tenths (*dekatai/oi/a*) away from the money that is being paid in.

NOTES ON READINGS

Line 1: This line was written on the fascia. The four letters are not evenly spaced. The dimensions are as follows, measured on centers: left edge to ⟨Θ⟩, 0.036 m.; ⟨Θ⟩ to Ε, 0.109 m.; Ε to Ο, 0.137 m.; Ο to Ι, 0.11 m.; Ι to right edge, 0.053 m. The dot in the first letter was not cut.

Line 2: The twenty letters of this line are arranged so as to occupy the full width of the stele, leaving margins of 0.009 m. at both left and right. They are larger than the letters in any other line on the stone. In the last letter-space most of the original surface is gone, but I think that I can detect traces of the angle formed by the two bottom strokes of sigma.

Line 3: I cannot see any traces of letters in stoichoi 30 and 31 at the end of this line. Although the twenty-nine letters in this line were inscribed in the same stoichedon arrangement as in the rest of the text from here to the bottom of the stele, the last two stoichoi were apparently left blank to make line end and word end coincide. [ἐκ] might be possible.

Line 4: To make the heading that labels the main body of the inscription stand out more prominently, the cutter left the surface after νήσων blank and inscribed the name of the proposer at the beginning of line 5 followed by the text of the law itself. For other examples of this device, see Osborne 1973, pages 260–264.

Line 8: Of the dotted kappa only a vertical survives at the left side of the stoichos.

Line 11: In the bottom, left corner of the thirtieth stoichos, there is the bottom of a vertical stroke.

Line 13: Of the dotted alpha I can make out traces of the two diagonal strokes, but the surface of the stone in between them has been lost.

Line 14: At the left side of the last preserved stoichos, there is part of a diagonal stroke sloping down from right to left.

Line 15: At the left side of the thirtieth stoichos, there is a vertical stroke with the surface to the right of it worn away.

Line 17: Of the dotted nu only the left vertical is preserved.

Line 21: In the worn area of the twenty-ninth stoichos, there is the outline of a triangular letter.

Line 22: At the left side of the twenty-ninth stoichos, part of a vertical stroke is visible.

Line 23: The cutter did not inscribe the crossbar of alpha in the third stoichos and the dot in the theta in the twenty-third.

Line 24: Of the initial alpha the cutter failed to inscribe the crossbar. The surface inside the circular letter in the twenty-ninth stoichos is not preserved.

Line 25: The cutter did not inscribe the crossbar of eta in the twentieth stoichos.

Line 26: Again he omitted the crossbar of eta in the third stoichos.

Line 29: In the thirteenth and twenty-first stoichoi, the crossbar of alpha was never inscribed.

Line 32: Of the twenty-eighth letter only a portion of the left diagonal survives at the left edge of the stoichos.

Line 34: The crossbar of the alpha in the eleventh space was omitted by the cutter.

Line 35: Of the twenty-seventh letter I can see only a diagonal stroke at the right side of the space sloping down from left to right.

Line 37: In the first two letter-spaces the cutter failed to inscribe the crossbar of the alphas.

Line 40: Of the dotted nu in the twenty-ninth stoichos, only a vertical stroke survives at the left side.

Line 46: The cutter added a superfluous stroke, resembling the top diagonal of kappa, to the iota in the twenty-seventh letter-space.

Line 48: Of the dotted alpha in the thirty-first stoichos, only the bottom of a diagonal stroke survives in the lower left corner.

Line 50: At the left side of the last stoichos there survives a vertical stroke.

Line 55: The cutter failed to inscribe the crossbar of the alpha in the fourth stoichos.

Line 58: The crossbar of eta in the eleventh space was never inscribed. For the disturbance in the stoichedon order in this line, see below, pages 11–12.

Line 60: The cutter never inscribed the crossbar of the alpha in the thirteenth space.

Line 61: The same is true of the alpha in the ninth stoichos.

LAYOUT

With only one exception (line 58), the stoichedon pattern of thirty-one letters in lines 5–61 was rigidly followed throughout. There are no blank spaces and no instances of two letters squeezed into one stoichos. Although the cutter sometimes omitted strokes, primarily the crossbar of alpha and eta, he never left out a letter, nor was he ever guilty of dittography. The lettering is very handsome. Despite the discoloration of the inscribed surface and the water damage along its right side, one can still appreciate the aesthetic appeal of this stele. With its smoothly polished, white background for regular rows of pleasing letters, probably painted red, this monument gives further proof that even when publishing complex administrative details the Athenians still had an eye for beauty.

The width of the stele tapers upward from 0.437 m. at the level of its final line (61) to 0.422 m. at the level of line 2. Despite the broadening of the field at his disposal as he moved down the stone, the cutter maintained a uniform margin at the left of 0.005–0.007 m. in width. The right side of the inscribed surface is too poorly preserved to measure the width of its margin, but the latter was probably symmetrical with that on the left. While avoiding the unpleasing effect of tapering margins, the cutter also chose not to occupy the additional space created by the stele's downward broadening by increasing the length of his stoichedon line to thirty-two letters. He was able to solve his spatial problems and create the remarkably regular appearance of this inscription by gradually expanding the horizontal checker-units. The result is that the thirty-one letters in each line in the lower portion of the inscription are slightly more widely spaced than their counterparts in the upper portion. This is not readily apparent to the reader. Only measurement reveals the stages by which the cutter made the transitions. The extremes are line 5 (the first full line of thirty-one letters), where eleven letters measured on centers equal 0.135 m., i.e., a checker-unit of 0.0135 m., and line 61, where the same number of letters occupy 0.141 m., a checker-unit of 0.0141 m. The former spacing seems to have been employed down to lines 17–20, where the cutter expanded to 0.137 m. for eleven letters, i.e., a checker-unit of 0.0137 m., which he then followed down to line 40. Here he expanded again to 0.139 m. for eleven letters, i.e., a checker-unit of 0.0139 m., until lines 60–61, in both of which eleven letters measured on centers occupy 0.141 m. He made his adjustments in the horizontal stoichedon spacing, therefore, at intervals of about twenty lines, i.e., roughly one-third and two-thirds of the way through the text of the law.

Nor did the cutter employ a static unit of vertical spacing. Again the extremes are, at the top, eleven lines measured on centers (lines 3–13) occupy 0.137 m., a vertical checker-unit of 0.0137 m., while at the bottom, lines 51–61, measured on centers, occupy 0.139 m., a vertical checker-unit of 0.0139 m. The difference is only slight, but clearly the cutter did increase the height of the interlinear space as he moved to the bottom of his text.

The only break in the strict stoichedon order of the text of the law occurs in line 58. The cutter's adjustment here was not abrupt, but subtle and barely perceptible. I confess that the first time I read the stone, when I was concentrating on construing its text, I did not notice this anomaly in the stoichedon order. Only when transcribing the text, letter-by-letter on squared paper, did I observe that there was something unusual about line 58. It has thirty-two letters, rather than the normal thirty-one, but one has to look twice to see this, since the cutter did not blatantly inscribe two letters in one stoichos. Rather, with omicron in the twenty-fourth stoichos, he began to shift each letter slightly out of alignment to the left, while still keeping each one roughly equidistant from its neighbors. This he did for the letters that would have stood in stoichoi 24 through 29. When he reached stoichos 29, having inscribed six letters in the preceding five stoichoi, he returned to the horizontal spacing he had used for the rest of the line.[3] That is, the last three letters in

[3] For examples of the more common expedient of inscribing three letters in two stoichoi, see Osborne 1973, pp. 267–268.

this line fall into their proper ranks just as if nothing unusual had happened to their companions to the left.

It is easier to describe than to explain the cutter's motives in placing thirty-two letters anomalously in line 58. Certainly, word or syllabic division, ignored elsewhere in the body of the law, was not the motive since the result is δυοῖν δεκάτ[.]|ιν. Nor was he preparing the way for further departures from the stoichedon pattern in subsequent lines, for they are all perfectly regular. One possibility is that when he reached stoichos 24, his eye dropped down one line in the copy he was using. The sequence of letters he had still to carve in the remaining eight stoichoi of line 58 runs as follows, ΟΙΝΔΕΚΑΤ. Directly below this in line 59 of his copy he would have seen ΟΙΚΗΣΙΝΚ. It is possible that he became confused at this point by the repetition of ΟΙ, ΟΙ in two consecutive lines and concluded, for some reason, that he needed to insert an extra letter into line 58. But, for all we know, other motives may have been at work. M. J. Osborne has well reminded us that stoichedon "inscriptions are the handiwork of fallible human masons, and it is no good expecting them to exhibit the regularity of a machine in their endeavours . . . the element of the unexpected is ever-present in stoichedon texts."[4]

Finally, we should ask why the cutter disposed his text in lines of thirty-one letters. Why not a round number such as thirty or an even one such as thirty-two? I suggest that after laying out his headings in lines 1–4, the cutter aimed at creating for the rest of his text a stoichedon inscription that would be as perfect as possible: that is, one in which, given the dimensions of the stele and the size of the letters, all the ranks would be full, leaving the fewest possible empty letter-spaces at the end. He came very close to doing this since only four stoichoi stand vacant at the end of line 61.[5] The text of the law in lines 5–61, excluding headings, contains a total of 1,764 letters. A stoichedon line of thirty letters would have required fifty-nine lines with seven empty spaces; a line of thirty-two letters yields fifty-five lines with twenty-eight vacats. Clearly the module of thirty-one letters produced fewer vacant spaces in the final line.

Contemporary parallels suggest that some cutters may in fact have made such calculations. The decree of Aristoteles outlining the terms of the Second Athenian Confederacy, 378/7 B.C., was also inscribed with a stoichedon line of thirty-one letters. It consists of seventy-two lines of which the last contains no blank spaces.[6] The mason of Nikophon's law on silver coinage, 375/4 B.C., filled every space in the fifty-four lines of the main text by employing a stoichedon line of thirty-nine letters, *SEG* XXVI 72. The main body of *IG* II^2 109, a citizenship decree for Astykrates of Delphi, 363/2 B.C., consists of fifty-five full stoichedon lines of thirty-one letters each, except for line 57, which comes before the rider. These and other examples of Attic decrees of this period in which the final stoichedon line contains no blank spaces may be the result of calculations similar to those I am attributing to the cutter of the grain-tax law.[7]

Such calculations, of course, assume that some cutters may have counted the letters of the text they were employed to copy before they actually began to inscribe the stone. Added to aesthetic concerns, there might have been a monetary motive for this procedure if, as was sometimes the case, cutters were paid by the letter. For a helpful discussion of a cutter's preparations before inscribing, see Tracy 1975, pages 115–120. For the master cutter Deinomachos at Delphi in the

[4] Osborne 1973, p. 259.

[5] Ironically, had the cutter not temporarily abandoned the stoichedon order in line 58, he would have been required to leave only three spaces blank at the end.

[6] *IG* II^2 43, lines 7–77. The parallel is not exact, since there are anomalies in the stoichedon order in lines 24, 45, and 69.

[7] Other examples include *IG* II^2 53 (*SEG* XXXII 45), before 387/6 B.C.; 133, 355/4 B.C.; 145 + *SEG* XIV 50 (*SEG* XXXIII 69), before 358/7 B.C.; and 148, ca. 356/5 B.C. Of the 229 decrees of the 5th century B.C. in *IG* I^3, the only example of a stoichedon inscription in which the last line contains no blank spaces is *IG* I^3 127 of 405/4 B.C., "stoich." 57–61. The last line of *IG* I^3 102, from 410/9 B.C., has only one *vacat*, stoich. 36. This phenomenon of stoichedon texts of ca. 405–350 B.C. containing no blank spaces in their final lines might repay closer investigation.

4th century B.C., who was paid at two different rates per 100 letters depending on their size, see Bousquet 1989, page 167 and his index, page 287. For other cutters outside of Attica who were paid on the basis of the number of letters, see, e.g., *IG* IV^2 1.108, line 169, Epidauros; *IG* VII 3073, lines 10–12, τῶν δὲ γραμμάτων| τῆς ἐγκολάψεως καὶ ἐγκαύσεως στατῆρα καὶ| τριώβολον τῶν χιλίων γραμμάτων, Lebadeia ("for the inscribing and the painting of the letters, one stater and three obols per 1,000 letters"); *IG* XI 159, line 66; 161A, lines 118–119; 199C, lines 70–80, Delos. This list is by no means complete.

Similar epigraphic evidence that cutters at Athens were compensated on the basis of the number of letters contained in a given document is apparently lacking. The cause was probably the Athenian practice of allotting fixed sums in ten-drachmai increments (ten to sixty dr.) for the cost of setting up inscribed state documents. The cutter's fee, which could indeed have been calculated on the basis of the number of letters in the text, was included in the sum allocated, but the presence of other costs such as the price of the stone, transport, lead for the plinth, labor for its erection, and so forth makes it impossible to determine the amount of this fee. See Nolan 1981.

Our cutter omitted to inscribe the horizontal bar of alpha eight times. Without wishing to enter the controversy as to whether or not Attic cutters drew their letters on the stone in paint or chalk before inscribing them, I do think that it is perhaps significant that in seven of the eight instances another alpha lies in the immediate vicinity, either in the very next stoichos or, at the most, no more than four letters away. This suggests that using one tool for the diagonals of alpha and another for the crossbars, he may have cut the former strokes first for two or more alphas in close proximity and then was not always careful to go back and cut the shorter crossbar in all of the alphas in question. For the possibility that this may also have happened sometimes with eta, see below, page 56.

Since this inscription is precisely dated and my commentary is already extensive, I have not searched exhaustively for other surviving Attic inscriptions that may have been cut by the same mason. Nor am I confident in my ability to attribute several different inscriptions to the same cutter in the manner that has been so profitably perfected by Stephen V. Tracy.[8] A list of inscriptions cut by the same man who inscribed our law of 374/3 B.C. might, in any case, tell us more about the others than about the text that now concerns us. For what it is worth, the lettering on the following inscriptions seems to my untutored eye fairly closely to resemble that on the grain-tax law: *IG* II^2 43; *SEG* XVII 20; XXVI 72; XXVIII 48.

[8] For impressively persuasive examples of his method and its results, see Tracy 1990 and 1995, with earlier bibliography.

CHAPTER II

COMMENTARY ON THE LAW

LINE 2

Ἐπὶ Σωκρατίδο ἄρχοντος

Sokratides is firmly dated as the *archon* of 374/3 B.C.; see Diodoros 15.41.1 and other testimonia in Develin 1989, page 243. In the absence of a full preamble, we do not have any explicit evidence as to the time of the year or the *prytany* when the Athenians passed this law. For speculation on this topic, see below, pages 72–73, 119–120.

More information, including the day of the *prytany* when the *nomothetai* voted approval of the law, was undoubtedly recorded on the copy in the archives. Diokles' law of 400 B.C. prescribed that τοὺς δὲ μετ' Εὐκλείδην τεθέντας καὶ τὸ λοιπὸν τιθεμένους κυρίους εἶναι ἀπὸ τῆς ἡμέρας ἧς ἕκαστος ἐτέθη, πλὴν εἴ τῳ προσγέγραπται χρόνος ὅντινα δεῖ ἄρχειν . . . τὸ δὲ λοιπόν, ὃς ἂν τυγχάνῃ γραμματεύων, προσγραφέτω παραχρῆμα τὸν νόμον κύριον εἶναι ἀπὸ τῆς ἡμέρας ἧς ἐτέθη (Demosthenes 24.42): "Laws passed after (the archonship of) Eukleides (403/2 B.C.) and laws passed in the future are to be valid from the day on which each one was passed, unless a law has added to it a clause stipulating the time of its inception. . . . For the future let each secretary who happens to be in office at the time add immediately the clause that the law is to be valid from the day on which it was passed."

Since the text of our law does not stipulate a specific time when it is to go into effect, we may infer that it was valid from the day (unknown to us) on which the *nomothetai* passed it. For useful discussion on this point, see Hansen 1990; West 1989.

LINES 3–4

Νόμος περὶ τῆς δωδεκάτης τοῦ σίτου *vv*
τῶν νήσων

Although the text inscribed on this stele lacks the enactment clause, ἔδοξεν/δεδόχθαι τοῖς νομοθέταις ("resolved by the *nomothetai*"), attested in other surviving Athenian laws on stone, the heading in these lines clearly establishes it as a *nomos* and not a *psephisma* of the *boule* and *demos*.[1] To make the heading more prominent, the cutter did not use larger letters or space them more widely. Observing word-division, he simply left vacant the last two stoichoi in line 3 and the remainder of line 4 after the last word (νήσων). No other such headings labeling the contents of an inscribed Athenian law have survived on stone. The wording is reminiscent of the decree of 409/8 B.C. which authorized the republication of τὸ[ν] Δράκοντος νόμον τὸμ περὶ τõ φό[ν]ο ("the law of Drakon on homicide"), *IG* I^3 104, lines 4–5; and the law of 353/2 B.C. with its reference to [κατὰ τὸ]ν Χαιρημονίδο νό[μον τὸν περὶ τῆ]ς ἀπαρχῆς ("according to the law of Chairemonides concerning the first-fruits"), *IG* II^2 140, lines 8–10, 32–34.

This is the ninth surviving Athenian *nomos* on stone from the 4th century B.C. The others, in chronological order, are (1) *SEG* XXVI 72, law on silver coinage, 375/4 B.C.; (2) unpublished; from the Agora Excavations, inv. no. I 7495, law concerning public finance, 354/3 B.C., see Alessandrì 1982, pages 7–11; (3) *IG* II^2 140, law on Eleusinian first-fruits, 353/2 B.C.; (4) *IG* II^2 244, law

[1] For the distinction both in terminology and in substance, see Hansen 1978b; Hansen 1979a; both updated in Hansen 1983, pp. 161–206. See also Hansen 1991, pp. 161–177, 256–257; Rhodes 1987; Sealey 1987, pp. 41–45.

on rebuilding the walls, 337/6 B.C.; (5) *SEG* XII 87, law against tyranny, 337/6 B.C.; (6) *IG* II^2 334 + *SEG* XVIII 13, law on the Lesser Panathenaia, 336–334 B.C.; (7) *IG* II^2 333, law regarding offerings, 335/4 B.C.; (8) *SEG* XXXV 83, law on the sanctuary of Artemis at Brauron, undated: 3rd century B.C.(?).

Related documents include the following three decrees that were referred to the *nomothetai: IG* II^2 222, honorary decree of Peisitheides of Delos, 344/3 B.C.; *IG* II^2 330, honorary decree for Phyleus of Oinoe, 335/4 B.C.; *IG* VII 4254 (*Syll.*[3] 298), honorary decree for the *epimeletai* of the Amphiareion, 329/8 B.C.

Hansen (1979a, pages 32–35) suggests that the following documents might also be classed as *nomoi: SEG* XXI 255 + 257 + 346 (XXX 61), regulations on the Eleusinian Mysteries, 367–348 B.C.; *IG* II^2 412 (*SEG* XXXII 81), regulations on the procedure of *phasis* ("ex lege haec supersunt," Kirchner), 336/5–322/1 B.C.; *SEG* XVI 55, regulations for a festival at Eleusis (?), 330/29 B.C.(?); *SEG* XXV 82, regulations for the Dipoleia and the Bouphonia, 4th century B.C.

Humphreys (1985, page 202) considers the inscription published as a decree in Schweigert 1940, pages 325–327, no. 35, to be "a new law concerning Lemnos."

Apart from the heading in lines 3–4, our new law contains no other information about the process of *nomothesia* and the procedure followed by the *nomothetai.* Lacking also are any instructions at the end regarding the publication of the law or payment for the stele. Although the law gives instructions to the newly elected ten men (lines 36–55), the *ekklesia* (lines 42–46), the *apodektai* (lines 55–61), and, by implication, the *poletai* (lines 6–10, 28–31), it is silent on the vexed question of the specific relationship between the *nomothetai* and the *boule* and the Assembly. For this problem, see Stroud 1974, pages 162–163; Humphreys 1985, pages 201–204.

Our law provides the earliest evidence for the δωδεκάτη, a tax of $8\frac{1}{3}\%$, on the grain of the islands. Clearly the lawgiver and the *nomothetai* promulgated a whole new set of regulations governing the collection of such a tax and the disposition of its proceeds, but did they now establish it for the first time or were they making legislative changes in the collection of a tax already in existence? The use of the definite article here and in lines 6 and 47, τὴν δωδεκάτην, does not supply a decisive solution.[2] We have to infer the answer to this question from the provisions of the law itself; see below, pages 25, 37–39, 79–80.

Σῖτος in this law refers exclusively to wheat, πυροί, and barley, κριθαί, both in the plural throughout, as lines 9–10, 21–24, 45 make clear; see below, pages 40–44, 54–56, 73–76, commentary on lines 8–10.[3] The islands in question are Lemnos, Imbros, and Skyros; see below, pages 31–37, commentary on line 7.

LINE 5

Ἀγύρριος εἶπεν

As in Nikophon's law on silver coinage of the previous year, *SEG* XXVI 72, the name of only one individual, the proposer, appears on the stone in addition to the name of the archon. Unlike

[2] Cf. how Thucydides (7.28.4) describes the institution of the 5% tax in the Athenian empire in 413 B.C., καὶ τὴν εἰκοστὴν ὑπὸ τοῦτον τὸν χρόνον τῶν κατὰ θάλασσαν ἀντὶ τοῦ φόρου τοῖς ὑπηκόοις ἐποίησαν ("and about this time they instituted the 5% tax on sea-borne commerce for their subjects in place of the tribute").

[3] For the wider sense of σῖτος to include other cereals or a restrictive sense limited to only one, see Jardé 1925, pp. 1–4. Happily, in our law we do not have to face the dilemma he poses on p. 4, "Quand un texte littéraire ou épigraphique porte σῖτος, comment faire le départ entre les diverses céréales, ou même simplement entre les deux principales, le blé et l'orge?" For discussion of the history of the word σῖτος, with helpful bibliography, see Gallo 1983, pp. 455, 462, note 15, 468–469, note 78, ably refuting the attempt of Moritz (1955, p. 136) to fix 327/6 B.C. as a *terminus ante quem* for σῖτος and πυρός becoming synonyms; useful also on this point are Gauthier 1976, p. 48; Amouretti 1979 and 1986, pp. 36–40; Battaglia 1989, pp. 39–51; Sallares 1991, pp. 347–348; Isager and Skydsgaard 1992, pp. 21–26.

the otherwise unknown Nikophon, however, the author of the grain-tax law may be plausibly identified. Despite the absence of patronymic and demotic, the name of the proposer is so rare as to limit identification to the only known homonymous contemporary Athenian, Agyrrhios I of Kollytos, Kirchner, *PA* 179; Davies 1971, no. 8157, I–II (stemma page 281); Osborne and Byrne 1994, page 8, *s.v.* Ἀγύρριος, no. 1; Traill 1994, no. 107660.[4]

Davies has persuasively urged that he was born at the latest by ca. 440 B.C. and not, in any case, earlier than ca. 450 B.C. These dates would mean that he was roughly sixty-six to seventy-six years old when he proposed the present law. We know of no certain activity of Agyrrhios later than our inscription of 374/3 B.C. See below, pages 23–25.

Although others have discussed his career in some detail,[5] I take it up again here since our new law casts several features of his biography in a different light. In turn, many of Agyrrhios' previous activities are significantly reflected in some of the provisions of his law of 374/3 B.C. Davies (1971, pages 278–279) suggested that his "family's origins and interests lay in business rather than in landholding [and his] politico-financial acts must be seen . . . against the fifth century vision of a participating democracy—a vision which he must be taken to have shared."

His father's name is unknown, but his deme is attested as Kollytos by Demosthenes 24.134; *IG* II^2 1, line 41; 2, line 1 [restored on the basis of line 6]; *SEG* XXVIII 45 [wholly restored, see below, page 19]. We cannot be sure that Agyrrhios actually resided in his own city deme, but it may be more than coincidence that the large Sanctuary of Aiakos, which is to serve as the repository for the grain from the islands in our law (lines 14–16), was located on the edge of the Agora in Kollytos (see below, pages 77, 88–90). A direct connection cannot be proved. It would be naive, however, to rule out a possible (profitable?) link between this large sanctuary, now transformed into a granary in Agyrrhios' own neighborhood, and his previous extensive experience in both business and public service.

In 405 B.C. Aristophanes' chorus of frogs included in a long list of those barred from their Bacchic revels (*Frogs* 367–368)

> ἢ τοὺς μισθοὺς τῶν ποιητῶν ῥήτωρ ὢν εἶτ' ἀποτρώγει,
> κωμῳδηθεὶς ἐν ταῖς πατρίοις τελεταῖς ταῖς τοῦ Διονύσου.
>
> Or a politician who goes so far as to nibble away at the poets' pay when
> he has been mocked in the ancestral rites of Dionysos.

The scholiasts offer two identifications of this unnamed politician: τοῦτο εἰς Ἀρχῖνον. μήποτε δὲ καὶ Ἀγύρριον. μέμνηται δὲ τούτων καὶ Πλάτων ἐν Σκευαῖς (*PCG* VII, F 141) καὶ Σαννυρίων ἐν Δανάῃ (*PCG* VII, F 9). οὗτοι γὰρ προιστάμενοι τῆς δημοσίας τραπέζης τὸν μισθὸν τῶν

[4] To my knowledge only one other certain Athenian Agyrrhios is so far known, Ἀγύρριος II Καλλιμέδοντος [Κολλυ]τεύς, Kirchner, *PA* 180; Davies 1971, no. 8157, III; Osborne and Byrne 1994, p. 8, *s.v.* Ἀγύρριος, no. 2; Traill 1994, no. 107665. Active ca. 285–282 B.C., he was probably the great-grandson of Agyrrhios I. For the fictive Ἀγύρριος ἐκ Λακιαδῶν of Kirchner, *PA* 181 and 12375, *IG* I^{2}639 = *IG* I^3 854, see Raubitschek 1949, pp. 227–229, no. 193; Traill 1994, no. [107670].

The gravestone *IG* II^2 10570, found in Athens, bears the simple inscription Ἀγύρριος, without patronymic, demotic, or ethnic. Kirchner excluded it from *PA*, but it is included by Osborne and Byrne (1994, p. 8, *s.v.* Ἀγύρριος, no. 3); Traill 1994, no. 107655. In view of the fact that it cannot be accurately dated, however, it could have belonged to Agyrrhios I or II. Had it been set up with other gravestones in a family burial plot, the absence of patronymic and demotic would not be decisive. The name Agyrrhios is even rarer outside of Attica.

[5] To the good bibliography in Davies 1971, p. 277 add Rhodes 1981, pp. 492–493; Sartori 1983; Strauss 1987, pp. 101–102, 130–132, 135–136, 142, 155–163; Rankin 1988, pp. 191–192; Hansen 1989, p. 34; Traill 1994, no. 107660. I have not consulted the unpublished Princeton dissertation of Hess (1963), in which Agyrrhios' "career is reconstructed . . . Hess . . . sees him (very unconvincingly) as 'the focal point of the political satire of the *Ecclesiazusae*' (a latter-day substitute for Cleon)" (Ussher 1973, p. 101). Ussher's assessment of Hess is shared by Rothwell (1990, pp. 5–6, 102).

I see nothing to recommend Dindorf's emendation of the mss. reading διώξει Σμικύθην καὶ κύριον to καὶ Ἀγύρριον in Aristophanes, *Knights* 969 of 424 B.C.; followed by van Leeuwen (1940, p. 171).

κωμῳδῶν ἐμείωσαν κωμῳδηθέντες.[6] "This refers to Archinos and possibly also to Agyrrhios. Platon mentions these men in his *Skeuai* and Sannyrion in his *Danae*, for these men, being in charge of the public bank, decreased the pay of the comic poets when they had been ridiculed." Agyrrhios is named alone in a scholion to Aristophanes, *Ekklesiazousai* 102 as responsible for reducing the *misthos* of the poets, ὁ Ἀγύρριος στρατηγὸς θηλυδριώδης, ἄρξας ἐν Λέσβῳ. καὶ τὸν μισθὸν δὲ τῶν ποιητῶν συνέτεμε (καὶ πρῶτος ἐκκλησιαστικὸν δέδωκεν).[7] "Agyrrhios: an effeminate general, having held office in Lesbos. He also reduced the pay of the poets (and was the first to give pay to the assembly)."

If we can trust this information, Agyrrhios was prominent enough in public life before 405 B.C. to have been mocked on stage by the comic poets and to have successfully joined the well-known politician Archinos in carrying through a measure to reduce their pay.[8] Whether the motive for this reduction was revenge, as Aristophanes charges, or economy in tough times—or a little of both—we cannot tell. For students of the grain-tax law, however, an important consideration about Agyrrhios is that his earliest known activity involved public finance. If Aristophanes' characterization of ῥήτωρ can apply to Agyrrhios, he may have been active in politics and a successful advocate in the *ekklesia* for reduced pay for poets.[9] If there is anything more than guesswork behind the scholiast's designation of Archinos and Agyrrhios as προιστάμενοι τῆς δημοσίας τραπέζης ("in charge of the public bank"), he may have even been operating in some kind of an official financial capacity.[10]

Agyrrhios' whereabouts and his political activity at the time of Aigospotamoi and the revolution of the Thirty are not attested in our surviving sources. He was certainly in Athens and a member of the *boule* in the archonship of Eukleides, 403/2 B.C., however, for the decree passed in the *prytany* of Pandionis of that year reaffirming citizenship for the *demos* of the Samians records him as *grammateus*, *IG* II^2 1, lines 41–42. His membership in this first council of the newly restored democracy is probably a good indication of his political sympathies, for he is not likely to have cleared his *dokimasia* for this post had he been tainted by any ties to the Thirty.[11] His secretaryship may also be a mark of his prominence and the high regard in which he was held by his fellow councillors, since at that time the secretary of the *boule* was elected by a show of hands of his colleagues.[12] The only other known *grammateus* from 403/2 B.C. was also a very prominent politician, Kephisophon I Paianieus, Kirchner, *PA* 8401 = 8400? = 8415? = 8416?; Davies 1971, page 148 D, who held office in the *prytany* of Erechtheis. It is clear from the two decrees in *IG* II^2 1, lines 41–55 and 55–75 that Agyrrhios, as secretary in Pandionis, preceded Kephisophon in Erechtheis. This cannot be construed as certain evidence that the former was more highly regarded by his colleagues than was Kephisophon, however, for the latter, as a tribesman of Pandionis, was excluded from the secretaryship while his own tribe held the *prytany*.

Agyrrhios has also been restored as the *grammateus* in two other inscribed documents attributed to the *prytany* of Pandionis in 403/2 B.C. The first of these is *IG* II^2 2, a proxeny decree that

[6] Dübner 1877, p. 286, line 367. The scholia of Tzetzes, who gives the name as Argyrios, repeat this information; see Koster 1962, pp. 803–804, 809, 858 on lines 367, 406, and 585, respectively.

[7] Dübner 1877, p. 316 on line 102, repeated in the *Souda Lexikon, s.v.* Ἀγύρριος, ed. Adler I, p. 39, with the variation ἄρξας ἐν Λήμνῳ; see below, p. 22.

[8] For the *misthos* of the comic poets, see Pickard-Cambridge 1968, p. 90.

[9] For the meaning "politician" here, see Hansen 1991, pp. 143–145; Dover 1993, p. 242.

[10] For discussion of this passage with bibliography, see Sartori 1983, pp. 56–62 and Bogaert 1968, pp. 81, 88–89. Bogaert does not believe that there was a state bank at Athens much before 329 B.C. and concludes, "Le scoliaste est influencé ici par des institutions de l'époque hellénistique et a commis un anachronisme." The anachronism may, however, be limited to the scholiast's terminology.

[11] Cf. Rhodes 1981, p. 492, "from the silence of And. I. *Myst.* 133–6 we may assume that he was on the democratic side in 404–403"; similarly Strauss 1987, p. 101. Sartori (1983, pp. 64–65), going beyond the evidence, sees Agyrrhios as "un continuatore di Cleofonte" and a "predecessore di Eubulo."

[12] Aristotle, *Ath* 54.3; Rhodes 1972, pp. 134–136; 1981, pp. 599–603.

Wilhelm restored as having been passed on the same day as *IG* II^2 1, lines 41–55.[13] Agyrrhios' name may have been written up in larger letters at the top of this stele, [Ἀγύρριος Κο]λλυτεὺς ἐγραμμάτευ[ε], line 1, in the manner described by Aristotle, *Ath* 54.3: καὶ γὰρ ἐν ταῖς στήλαις πρὸς ταῖς συμμαχίαις καὶ προξενίαις καὶ πολιτείαις οὗτος (ὁ γραμματεύς) ἀναγράφεται, "Indeed on the stelai dealing with alliances, *proxenia*, and citizenship, (the name) of (the secretary) is written at the top." The second inscription is the large and beautiful marble stele that once contained a catalogue, an epigram, and a decree in honor of those who captured Phyle and brought back the *demos* to Athens, *SEG* XXVIII 45. Although the first *prytany* of 403/2 B.C. would be an appropriate time to pass such a decree, we have to remember that not one letter of Agyrrhios' name is preserved on the stone. Since a total of only ten possible letters survives from the first four lines of the presumed decree, Raubitschek's restoration of a stoichedon line of eighty-one letters cannot be taken seriously as evidence that Agyrrhios was secretary when it was passed.[14]

Agyrrhios' next recorded activity is directly relevant to his authorship of the grain-tax law. Probably in 402/1 B.C.,[15] according to Andokides 1 (*On the Mysteries*) 133–134, Agyrrhios was himself the leader of a group of men who farmed the 2% tax at Athens, ἀρχώνης ἐγένετο τῆς πεντηκοστῆς.[16] At this time, he successfully bid thirty talents for the right to collect this tax for one year and took as his associates a group of men characterized by Andokides as all those who gather under the poplar tree, ἐπρίατο τριάκοντα ταλάντων, μετέσχον δ' αὐτῷ οὗτοι πάντες οἱ παρασυλλεγέντες ὑπὸ τὴν λεύκην. In his defense speech *On the Mysteries*, Andokides accused this group of conspiring to keep the bidding for the *pentekoste* at a low level, with the result that they turned a profit of three talents in their first year, 402/1 B.C. When Agyrrhios and his associates again allegedly conspired to repeat their bids for this tax at the auction in the following year, 401/0 B.C., at the same level of thirty talents, Andokides claims to have routed them by bidding thirty-six talents. He asserts that he and his associates were able to turn a small profit by collecting more than this sum, thereby preventing his opponents from enjoying revenues that rightfully belonged to the Athenians: συνέστησαν πάντες καὶ μεταδόντες τοῖς ἄλλοις ἐωνοῦντο πάλιν τριάκοντα ταλάντων. ἐπεὶ δ' οὐκ ἀντωνεῖτο οὐδείς, παρελθὼν ἐγὼ εἰς τὴν βουλὴν ὑπερέβαλλον, ἕως ἐπριάμην ἓξ καὶ τριάκοντα ταλάντων. ἀπελάσας δὲ τούτους καὶ καταστήσας ὑμῖν ἐγγυητὰς ἐξέλεξα τὰ χρήματα καὶ κατέβαλον τῇ πόλει, καὶ αὐτὸς οὐκ ἐζημιώθην, ἀλλὰ καὶ βραχέα ἀπεκερδαίνομεν οἱ μετασχόντες· τούτους δ' ἐποίησα τῶν ὑμετέρων μὴ διανείμασθαι ἓξ τάλαντα ἀργυρίου, "They all conspired together and sharing it out among the others, they were trying to buy (the right to collect the tax) for thirty talents. But when no one was offering a counter bid, I came forward in the *Boule* and kept raising the bid until I bought (the right to collect the tax) for thirty-six talents. Having driven these men away and provided you with guarantors, I collected the money and paid it to the *polis* and I did not myself suffer any loss, but my associates and I were actually able to realize a small profit, while I kept these men from distributing among themselves six talents of money that belonged to you."

Apart from its useful verbal parallels with some parts of our new law, this passage reveals that at least as early as 402 B.C. Agyrrhios was probably influential enough in the business community

[13] *Pace* R. Develin (1989, p. 227), Wilhelm's restoration of this decree seems to me more plausible than that offered by Walbank (1982b), who places it in 382/1 B.C. D. M. Lewis also preferred Wilhelm's dating and restorations; see *SEG* XXXII 38.

[14] Raubitschek 1941: "The restoration is naturally uncertain, but it so happens that the required space is exactly filled by a prescript that would date the proposal in the same year and the same prytany as I. G., II^2, 1, lines 41ff." (p. 295). Raubitschek's "required space" is in fact the result of his own restoration of lines 1–2, on which in turn he bases his reconstruction of the original width of the stone. But only two letters survive in line 1 and the evidence for determining the width of the stele and of the margins at the level of the decree is insubstantial.

[15] For the date, see MacDowell 1962, pp. 204–205.

[16] This tax is probably the 2% levy on imports and exports attested, e.g., in [Demosthenes] 34.7 and 35.29–30, cf. Hansen 1991, pp. 260–261; but D. M. Lewis' observation that many different Athenian taxes were called *pentekoste* makes such an identification uncertain (1959, pp. 243–244). See below, note 42.

to head a group of tax-farmers capable of winning the contract from the state and collecting a large sum of money. We cannot be certain that he was guilty of the charges of collusion leveled at him by Andokides,[17] but he seems clearly to have had direct personal experience as an entrepreneur with the intricacies of Athenian tax-farming. No doubt he put that experience to use in drafting the grain-tax law nearly three decades later. Andokides attributes to this incident the enmity that led Agyrrhios in 400 B.C. to join Kallias, Kephisios, and others in prosecuting him for impiety, 1.135–136.[18] This celebrated trial, which elicited Andokides' defense speech, *On the Mysteries*, again brought Agyrrhios very much into public prominence.

In the decade following the restoration of democracy, Agyrrhios took the political action for which modern scholars best remember him: the introduction of pay, *misthos*, for attending the *ekklesia*. Agyrrhios' initial stipend of one obol per meeting was doubled on the proposal of the naturalized Athenian Herakleides of Klazomenai. Later, sometime before the production of Aristophanes' *Ekklesiazousai*, ca. 393–390 B.C., Agyrrhios had raised the level of pay again to three obols.[19] The evidence is too slim for productive speculation on the motives for Agyrrhios' proposals,[20] the tactics he employed in winning public support for them,[21] or even the sources of the funds to pay for

[17] Sealey's warning (1956, p. 182), "It would be foolish to accept in every detail Andocides's account of the two quarrels, since no account from the other side is extant; but he is doubtless right in saying that Callias and Agyrrhios brought about the prosecution because of the quarrels," was not heeded by Buchanan (1962, p. 25), who accepted "the unpalatable story" at face value.

[18] For the possible political dimensions of this trial, see Sealey 1956, pp. 181–182; MacDowell 1962, pp. 10–18; Strauss 1987, p. 102; Missiou 1992, pp. 49–54. Contrast Funke (1980, pp. 22, note 15; 24, note 25; 116, note 44), who is skeptical.

[19] Aristophanes, *Ekklesiazousai* 186–188, 289–311; schol. on lines 102, 392; Aristotle, *Ath* 41.3. The exact dates of the introduction of pay by Agyrrhios and the two later increases are not known. See Hignett 1952, pp. 396–397, "The whole evolution described in *A. P.* 41.3 must have occurred within three years, for in view of the financial difficulties of Athens in and after 403 it is unlikely that pay for attendance at the assembly could have been introduced even at the rate of 1 obol until 395." Rhodes 1981, p. 292, "The introduction . . . must be dated soon after the democratic revolution." Ober 1989b, p. 98, "passed some time between 403 and 399." Aristotle, *loc. cit.*, implies that after the restoration of democracy there was an interval during which other devices were used to attract citizens to the assembly, μισθοφόρον δ' ἐκκλησίαν τὸ μὲν πρῶτον ἀπέγνωσαν ποιεῖν· οὐ συλλεγομένων δ' εἰς τὴν ἐκκλησίαν, ἀλλὰ πολλὰ σοφιζομένων τῶν πρυτάνεων ὅπως προσιστῆται τὸ πλῆθος πρὸς τὴν ἐπικύρωσιν τῆς χειροτονίας ("At first they decided against pay for the Assembly, but when the people were not attending the Assembly and the *prytaneis* were devising many ways of inducing enough people to attend to ratify the voting"). It was only after this that Agyrrhios proposed pay.

For the dating of the *Ekklesiazousai*, see the helpful bibliography in Ussher 1973, pp. xx–xxv (presenting the strongest case for 393 B.C.); Funke 1980, pp. 168–171; Strauss 1987, pp. 143, 149; Rothwell 1990, p. 2.

[20] Aristotle's motive still seems to me best; see note 19 and cf. *Politics* 4.1297a35–38; see also Strauss 1987, p. 101. For the moralistic approach, cf. Buchanan 1962, pp. 25–26, "We cannot exonerate Agyrrhios from sheer demagogy and lack of principle, . . . I can attribute it to no motive other than personal aggrandizement." In rejecting Aristotle, Ober (1989a, p. 326) argues (without supporting evidence) "that the reform occurred because the Athenian lower classes were determined to prevent the domination of the Assembly by the upper classes, who were seen (rightly or not) as likely supporters of oligarchy." I know of no ancient evidence to recommend this view or that of Burke (1992, pp. 223–224), to the effect that the *ekklesiastikon* "was meant to help subsidize the income of some of those for whom service as rowers [in the navy] was no longer regularly available." The proposal of Hansen (1986) that pay was intended to assure a quorum of 6,000 at the Assembly, was rejected by Gauthier (1990, pp. 439–441), who argued on the basis of Aristophanes, *Ekklesiazousai* and *I. Iasos* 20 that "le *misthos* était conçu pour assurer la ponctualité des citoyens; et c'était seulement par voie de conséquence, les Assemblées étant matinales et courtes, donc peu accaparantes, qu' il favorisait une forte participation civique." See also Gauthier 1993. For the opposite view see briefly Todd 1990, pp. 154–156. In the light of Gauthier's work, Hansen argued in 1996 that "the two objectives, to get more citizens to attend and to make them arrive earlier, are in no way mutually exclusive" (Forsén and Stanton 1996, pp. 30–33). On the adequacy of assembly pay and other aspects of this issue, see Markle 1985.

[21] Sinclair (1988, pp. 116–117) suggests that political competition between Agyrrhios and Herakleides drove up the level of the *ekklesiastikon*; cf. Funke 1980, p. 24, note 24; Strauss 1987, p. 101. Sealey (1956, p. 181) strangely opined that by 395 B.C. Herakleides had thrown Agyrrhios "into the shade by raising ecclesiastic pay to two obols." On the date and purpose of the honorary decree for Herakleides of Klazomenai, *IG* I^3 227 + II^2 65, see Funke 1980, p. 64, note 52, and the exchange between H. B. Mattingly and M. B. Walbank as reported in *SEG* XXXIX 9.

them.[22] Emerging clearly from Aristophanes' attacks on him in the *Ekklesiazousai*, however, is the prominence of Agyrrhios as a public figure in the 390s. Despite his alleged effeminate appearance in the past, he now wears a bushy beard—like the flute-player, Pronomos—and Aristophanes says that he πράττει τὰ μέγιστ' ἐν τῇ πόλει ("he manages the most important affairs in the city"), line 102, with scholia. While punning (inevitably) on his name (ἀργύριον, line 186) and reviling him as πονηρός for introducing *misthos* for the assembly, Praxagora nevertheless concedes that those who are paid adore him (lines 183–188).

> ἐκκλησίαισιν ἦν ὅτ' οὐκ ἐχρώμεθα
> οὐδὲν τὸ παράπαν· ἀλλὰ τόν γ' Ἀγύρριον
> πονηρὸν ἡγούμεσθα. νῦν δὲ χρωμένων
> ὁ μὲν λαβὼν ἀργύριον ὑπερεπῄνεσεν,
> ὁ δ' οὐ λαβὼν εἶναι θανάτου φήσ' ἀξίους
> τοὺς μισθοφορεῖν ζητοῦντας ἐν τἠκκλησίᾳ.
>
> There was a time when we had nothing whatever to do
> with Assemblies, but Agyrrhios at least we knew was a rascal.
> But now when we attend, the man who gets his money praises
> him to the skies, while the one who doesn't says that those who
> seek their wage in the Assembly ought to be put to death.

Later, the women of the chorus, on their way to the Pnyx in the early morning, lyrically anticipate their *triobolon*, lines 289–310. In the late 390s, then, Assembly pay and its main public advocate were still controversial topics at Athens.

Another roughly contemporary venture in the arena of public finance is attributed to Agyrrhios by Harpokration, who assigns to him the origin of the *theorikon*.[23] This is not the place to review the intricate modern controversy over the introduction of subsidized admission to the theater and its development into the Board for the Theorikon, which was probably instituted by Euboulos ca. 355 B.C., after the war of the Allies.[24] The arguments for rejecting the testimony of Harpokration, however, are not entirely persuasive. It still seems to me possible to interpret the ancient evidence as indicating that Agyrrhios introduced or reorganized subsidies for the theater in the archonship of Diophantos, 395/4 B.C., while recognizing Euboulos' transforming impact on this institution.[25] This move would be in keeping with Agyrrhios' introduction of Assembly pay and with Demosthenes' later characterization of Agyrrhios as δημοτικὸν καὶ περὶ

[22] Sealey 1956, p. 183: "Agyrrhios seems to have attached himself to Conon when the latter came back to Athens. It is not unlikely that it was the money brought by Conon that enabled Agyrrhios to raise ecclesiastic pay to three obols." Cf. Pritchett 1974, p. 119; Rothwell 1990, pp. 5, 12. Strauss (1987, pp. 130–131, 135–136), while rightly stressing that the evidence for a tie between Konon and Agyrrhios is unsatisfactory, finds the suggestion "attractive but unproven." Funke (1980, pp. 116–117) rejects it.

[23] *S.v.* θεωρικά, partly based upon Philochoros, *FGrH* 328 F 33.

[24] The issues are well laid out by Buchanan (1962, pp. 48–93, esp. 48–53), who accepts the attribution to Agyrrhios, as does F. Jacoby (*FGrH* III B, Supplement pp. 318–320); cf. Pickard-Cambridge 1968, pp. 265–268; Funke 1980, p. 24, note 24; Sartori 1983, pp. 62, 77. For a clear statement of the opposing view, i.e., rejecting the attribution to Agyrrhios, see Cawkwell 1963a, pp. 54–65; Ruschenbusch 1979, pp. 303–308; Rhodes 1981, pp. 492, 514; Sealey 1993, p. 256. Rich in bibliography on Euboulos and the *theorikon* is Leppin 1995.

In seeking a date for its origin, Pritchett's reminder (1991, p. 460, note 674), that the expenditure of the theoric fund "was essentially religious in character, incurred for various festivals devoted to the honor of the gods" should be borne in mind. For arguments in support of a Periklean origin of the *theorikon*, see Fornara and Samons 1991, pp. 72–75.

[25] This is achieved by taking Harpokration, *s.v.* θεωρικά; Libanios, *Hypothesis* 4 to Dem. 1; Hesychios, *s.v.* δραχμὴ χαλαζῶσα (= *Souda Lex.* Δ 1491); Zenobios, *Prov.* 3.27, ἐπὶ Διοφάντου τὸ θεωρικὸν ἐγένετο δραχμή ("under Diophantos the theorikon became a drachma") together to refer to the same event. It seems more likely to me that the last passage contains a reference to an archon date rather than to the regime of Diophantos of Sphettos (Kirchner, *PA* 4438), "the shadowy coeval of Euboulos," as suggested by Beloch (1923, III.1, p. 344), and followed by Ruschenbusch, Cawkwell, and Rhodes as cited in note 24. The happy quotation is from D. M. Lewis in Murray and Price 1990, p. 258.

τὸ πλῆθος τὸ ὑμέτερον πολλὰ σπουδάσαντα ("a man of the people and a zealous supporter of your Assembly," 24.134).

In or around 393 B.C., Agyrrhios appeared in court as a witness for the prosecution in the trial of the prominent banker Pasion. If we can trust the Pontic merchant for whom Isokrates wrote his *Trapezitikos*, Agyrrhios' status in the business community was such that Pasion had earlier enlisted his aid in trying to reach a settlement with the plaintiff in their financial dispute.[26] In this trial Agyrrhios' support for a man who had imported large quantities of grain to Athens may provide a hint of later events. We may be able to conclude not only that Agyrrhios had been concerned with the food supply considerably earlier than his law of 374/3 B.C. but clearly also that one of the most popular Athenian politicians of his day was working on intimate terms with the leading banker in Athens and with a very rich metic. Note that he did not carry out these activities behind the scenes but in the *dikasterion* in plain view of his fellow citizens (cf. Isokrates 17.3–4; 57).

The only securely dated evidence for Agyrrhios' military activity belongs in the summer of 389 B.C., when the Athenians sent him out as general to replace Thrasyboulos. The latter, after winning a battle over the Spartans and their Lesbian allies, had led his fleet from this island to Aspendos, where he was killed (Xenophon, *HG* 4.8.30–31; Lysias 28.5–12, 29; Diodoros 14.99.4–5). Our sources do not say what Agyrrhios did with the Athenian fleet once he took command. Xenophon drops the subject completely. Diodoros observes that after Thrasyboulos' death, the Athenian trierarchs quickly sailed from Aspendos to Rhodes; Diodoros implies that Agyrrhios found them there. It is possible that he simply brought the fleet back to Athens.[27]

If these passages were the only source of the formulation στρατηγὸς θηλυδριώδης ἄρξας ἐν Λέσβῳ ("an effeminate general who held office in Lesbos"), by which Agyrrhios is characterized in the scholia to Aristophanes, *Ekklesiazousai* 102 and *Ploutos* 176, then placing his command in Lesbos may be an incorrect inference.[28] Unless the scholiasts had other information on Agyrrhios' activities in 389 B.C., it is worth considering the variant reading—almost a *lectio difficilior*—ἐν Λήμνῳ, which is found in the *Souda Lexikon, s.v.* Ἀγύρριος. The reference then might be to another occasion on which Agyrrhios was general and posted to this fertile island. We must not lean too heavily on the slender reed of a variant reading in the *Souda*, but it is possible that Agyrrhios was general on more than one occasion. Firsthand knowledge of the agricultural potential of Lemnos would have been useful for the proposer of our grain-tax law.

His military activity in 389 B.C. has also been seen as the origin of an obscure passage in Plutarch (*Moralia* 801A–B), who quotes the following two lines of an unknown and undated comedy of Platon (*PCG* VII, F 201) spoken by Demos:

> λαβοῦ λαβοῦ τῆς χειρὸς ὡς τάχιστα μου·
> μέλλω στρατηγὸν χειροτονεῖν Ἀγύρριον
>
> Take, take my hand as fast as you can. I'm about to raise it to elect Agyrrhios general.

[26] Isokrates 17.31–32, where the speaker calls Agyrrhios ὄντ' ἀμφοτέροις ἡμῖν ἐπιτήδειον ("a close associate of both of us"); see Bongenaar 1933, pp. 146–148. On the date and occasion of this speech, see Trevett 1990. It is possible that Agyrrhios, the citizen, and Pasion, the metic, may have had a working relationship similar to the arrangements discussed by Millett (1991, pp. 224–229) and Cohen (1992, pp. 98–101). Strauss (1987, p. 142) aptly observes that Agyrrhios' nephew, Kallistratos, was also an associate (ἐπιτήδειος) of Pasion, [Demosthenes] 49.47.

[27] *Pace* R. Seager (*CAH*² VI, p. 115), Agyrrhios is unlikely to have taken the ships back to Lesbos or up to the Hellespont. It was Iphikrates whom the Athenians sent out in the same year with eight ships to carry on Thrasyboulos' projects in the Hellespont. No ships from Aspendos are said to have joined him (Xenophon, *HG* 4.8.34–39). For speculation about enmity between Agyrrhios and Thrasyboulos and the circumstances in which the former replaced the latter, see Strauss 1987, p. 102. This possibility is rejected by Funke (1980, pp. 159–161). On Thrasyboulos' campaign see Pritchett 1991, pp. 391–394, 488–489; Tuplin 1993, pp. 81–82.

[28] The same inference was drawn by Judeich (1893, *s.v.* Agyrrhios) and Edmonds (1957, pp. 548–549).

Although the passage provides additional evidence that Agyrrhios was a favorite target of the comic poets, there are no compelling grounds for associating it with his *strategia* in 389 B.C.[29] It shows only that at some unknown time in an unknown context the character Demos wanted to be rescued from voting for Agyrrhios as *strategos* in a comic play.

Davies (1971, pages 278–279) was reluctant to place Agyrrhios among the πλούσιοι because he found no evidence for liturgical service in the family. I am not aware of any new evidence for liturgies that has emerged since 1971. The members of Aristophanes' audience in 388 B.C., however, may have recognized him immediately as a rich man. In the *Ploutos*, Chremylos asks, "Is it not because of Wealth that the Great King has long hair?—Is it not because of Wealth that Agyrrhios farts?" Ἀγύρριος δ' οὐχὶ διὰ τοῦτον (i.e., Πλοῦτον) πέρδεται; line 176.[30] It is only a one-liner, but the audience was expected to get the point quickly. Agyrrhios' flatulence is due to his wealth.

The last known episode in Agyrrhios' career prior to the discovery of the grain-tax law of 374/3 B.C. is very briefly cited by Demosthenes to exemplify a powerful figure who was imprisoned for public debt, 24.134–135. The orator's characterization of him as an ἄνδρα χρηστὸν καὶ δημοτικὸν καὶ περὶ τὸ πλῆθος τὸ ὑμέτερον πολλὰ σπουδάσαντα ("a good man, a man of the people, and a zealous supporter of your Assembly") finds some echoes in what we know of his political career.[31] We shall see that it also aptly describes several features of our new law. Nevertheless, at some unspecified time he spent many years (πολλὰ ἔτη) in prison until he paid back the public money he had illegally retained. Demosthenes notes that his nephew, Kallistratos, who was a powerful figure at the time, did not resort to the device of proposing new laws in order to win Agyrrhios' release from prison.[32] Since no other source mentions this episode, we do not know how and when Agyrrhios incurred his public debt, how many years he remained in prison, and when he was released. Aristophanes, *Ploutos*, line 176 probably supplies a *terminus post quem* of 388 B.C.[33] Kallistratos' career is of little help in dating Agyrrhios' confinement more exactly, although Davies (1971, page 278) has suggested that "the burden of such a public debt might help to explain both Kallistratos' apparent obscurity in the 380s and the total obscurity of Agyrrhios' children, as well as the absence of the latter's descendants from the liturgical class."

Agyrrhios' successful authorship of our new law in 374/3 B.C. now raises the question whether he served his jail sentence in the period ca. 388–374 B.C. or at some date later than the passage of

[29] Edmonds (1957, pp. 548–549) not only makes this association but then uses it to date Platon's play Αἱ ἀφ' ἱερῶν to March 388 B.C. For the suggestion that Agyrrhios may have been general before 389 B.C., see Seager 1967, p. 113; Funke 1980, p. 161, note 107.

[30] The scholiasts on this line explain Agyrrhios' flatulence as a symptom of his great wealth. This passage is omitted from Traill 1994.

[31] Buchanan (1962, pp. 25–26, 52–53) rejects this assessment, preferring to see Agyrrhios as an opportunistic demagogue who pandered to the people for his own political ends, like the successors of Kleophon, cf. Aristotle, *Ath* 28.4 and Sartori, above, note 11.

[32] Sealey (1956, p. 187) speculates that Kallistratos now abandoned his old political associates and sought new ones: "This hypothesis explains why he allowed his uncle Agyrrhios to remain in prison for many years."

[33] Davies (1971, p. 278) proposes that "Agyrrhios suffered a fate similar to that of Pamphilos of Keiriadai . . . and about the same time." Pamphilos, a *strategos* in Aigina in 389/8 B.C., was prosecuted for embezzlement soon thereafter; his estate was confiscated and sold. Sources in Davies 1971, p. 365. Sealey (1956, p. 186) observes: "Agyrrhios was in disgrace; for it was probably now [387/6 B.C.] that he was imprisoned as a public debtor. If he had been the leading Athenian in the last years of the war, it is understandable that the Athenians turned against him when the war reached its unsuccessful conclusion." Cf. Strauss 1987, p. 161. The fact that Aristophanes represents Pamphilos and his associate in tears while in the next line Agyrrhios happily farts away may suggest that by 388 B.C. the latter had not yet shared the former's fate (*Ploutos*, lines 174–176). Pritchett (1991, pp. 489–490) aptly cites the slanders of bribery or extortion παρὰ τῶν ναυκλήρων καὶ ἐμπόρων ("from the shipowners and merchants") brought against the *strategos* Diotimos in 388 B.C. (Lysias 19.50) as setting "the tone for charges brought against all generals who had gained a name for wealth." There is no explicit evidence, however, for associating Agyrrhios' imprisonment with his political or military activity, as opposed to something like a failed business transaction involving state funds.

the regulations on the grain-tax. The former alternative would mean that he eventually emerged from prison to regain enough political influence to carry his law in the *nomothetai*. The latter alternative requires a long period of unattested inactivity, followed by the law of 374/3 B.C., then his conviction for misuse of public funds, followed by many years in jail and eventual release.

It is clear from Demosthenes 24.135, καὶ ἐγένετο ἐν τῷ οἰκήματι τούτῳ πολλὰ ἔτη, ἕως τὰ χρήματα ἀπέτεισεν ἃ ἔδοξε τῆς πόλεως ὄντα ἔχειν ("and he remained in this building [the prison] for many years until he paid back the money he possessed which had been decreed to belong to the *polis*"), that Agyrrhios did not die a public debtor in prison, as did for instance Pamphilos of Keiriadai (Davies 1971, page 365). He discharged his debt and emerged from the *desmoterion*. That he could then have returned to public life to propose the grain-tax law is not as improbable as it might seem at first glance. In this same passage of his speech *Against Timarchos*, Demosthenes parades before the *dikastai* several "just men" (δίκαιοι) who were nevertheless convicted of various offences and punished with incarceration. Among them is Thrasyboulos of Kollytos, who was twice imprisoned (134) and yet came out to serve as one of the Athenians' most trusted envoys to Thebes; see Aischines 3.138; Kirchner, *PA* 7305; Schwahn 1936, *s.v.* Thrasybulos no. 5.

On this reconstruction Agyrrhios' public career did not end with his long prison term. Some time after his release, he returned to legislative action with an elaborate proposal aimed at providing a supply of grain for the people of Athens and raising public revenues. In framing this law he drew upon his considerable practical experience as a tax-farmer, a prominent figure in the business community, and a politician skilled in public finance.

If, on the other hand, Agyrrhios' conviction for misuse of public funds followed the passage of his grain-tax law, we must find time for it and for many years in jail, followed by his eventual release, in the waning years of his career. As we have seen, by 374 B.C. he would have been ca. sixty-six to seventy-six years of age. There might be some attraction in following this chronology, as does Traill (1994, no. 107660), since it could permit the scenario that the occasion for the old entrepreneur's criminal act might have arisen in the climate of financial speculation that his own tax-law probably created.

I have given so much attention to the career of Agyrrhios because it provides a clearly and richly documented example of an Athenian citizen, prominent and very active in political life, elected in fact at least once to the highest office in the land, who at the same time openly engaged in business as a tax-collector with success enough to make him a very rich man. He obviously moved easily in the realms of the *boule*, the *ekklesia*, the *nomothetai*, and the Athenian military establishment. He was no stranger to the *dikasteria* and indeed to the *desmoterion*. Trusted by bankers and merchants as a reliable agent who was ready to use his influence in court in their behalf, he was familiar enough to comic audiences that the most successful playwright of his day could attack him in only a few short lines, confident that Agyrrhios would be recognized. It is difficult also to imagine that the framer of the present law on the grain-tax in the islands was not comfortable in what Julie Velissaropoulos has happily called "the world of the emporium."[34] We shall see revealed in his law knowledge of the carrying capacity of grain ships, the proportion of wheat to barley harvested on the islands, details of the weights and quality control necessary to produce the purest grain, familiarity with the mechanics of weighing and measuring it, how joint investment groups or *symmoriai* are put together, the yield of some of last year's taxes, and other details of Athenian public finance. Even if Agyrrhios did not acquire all of this knowledge at first hand—and there is no reason to conclude that he did not—it is a plausible inference that he knew *emporoi* and *naukleroi*,

[34] Velissaropoulos 1977. See also Bresson 1993. For spatial and commercial differences between *agora* and *emporion*, see Gofas 1993, pp. 167–195.

Athenian, metic, and foreign alike, intimately enough to frame and persuade his fellow *nomothetai* to pass this legislation that was designed to appeal to their entrepreneurial spirit while bringing conspicuous benefit to the *demos* and additional income to the military fund.

It will be well to bear Agyrrhios' career in mind when we are asked by modern scholars, as we sometimes are, to believe that trade and commerce in Athens were the preserve of the metic, that prominent politicians avoided the allegedly shady associations of the *emporion* and the Peiraieus, that "it was rare, to judge from the literary sources, for an Athenian citizen to engage in trade (i.e. overseas commerce)" (Hopper 1979, page 109; for similar sentiments, see, e.g., Erxleben 1974; Rickman 1980, pages 27–28; Finley 1985, pages 60, 144–145).

As long ago as 1975, on the basis of the literary sources, Isager and Hansen wrote, "Of all the *naukleroi* and *emporoi* who are mentioned in forensic speeches, fourteen are metics or aliens as opposed to fifteen citizens. The picture of Athenian foreign commerce must be given more nuances: metics and aliens were perhaps in the majority among *naukleroi* and *emporoi*, but they certainly did not dominate Athenian foreign trade completely" (page 72). For some of the advantages metics might have had over citizens in the wholesale grain business at Athens, see Montgomery 1986; Figueira 1986, pages 168–171. Especially persuasive on the substantial role of Athenian citizens in trade and commerce are W. E. Thompson 1982; M. V. Hansen 1984; and Montgomery 1986. It is well to bear in mind the words of the Athenian citizen who was a defendant in [Demosthenes] 33.4: ἐγὼ γὰρ . . . πολὺν ἤδη χρόνον ἐπὶ τῆς ἐργασίας ὢν τῆς κατὰ θάλατταν, μέχρι μέν τινος αὐτὸς ἐκινδύνευον, οὔπω δ' ἔτη ἐστὶν ἑπτά, ἀφ' οὗ τὸ μὲν πλεῖν καταλέλυκα, μέτρια δ' ἔχων τούτοις πειρῶμαι ναυτικοῖς ἐργάζεσθαι. διὰ δὲ τὸ ἀφῖχθαι πολλαχόσε καὶ διὰ τὸ εἶναί μοι τὰς διατριβὰς περὶ τὸ ἐμπόριον, γνωρίμως ἔχω τοῖς πλείστοις τῶν πλεόντων τὴν θάλατταν, "I have now for a long time been involved in maritime business and until a certain point I put myself at personal risk (on the sea), but it is almost seven years since I abandoned sailing, and possessing a modest amount of money, I try to put it to work on maritime loans. Because I have traveled to many places and have spent a lot of time in the *emporion*, I have come to be acquainted with most of those who make sea voyages."

LINES 5–6

ὅπως ἂν τῶι δήμωι σῖ[το]|ς ἦι ἐν τῶι κοινῶι

In order that there may be grain for the people in the public domain

The text of the law begins with a purpose clause, as in the Athenian laws recorded in *IG* II^2 334 + *SEG* XVIII 13, line 5; *IG* II^2 333, line 14; *SEG* XXXV 83, line 2; and the unpublished Agora I 7495, lines 4–6.[35] For the phraseology, cf. the decree of Kephisophon authorizing the colony to Hadria, 325/4 B.C., *IG* II^2 1629, lines 217–220, ὅπως δ' ἂν ὑπάρχηι| [τῶ]ι δήμωι εἲς τὸν ἅπαντα| [χρ]όνον ἐμπορία οἰκεία καὶ| [σι]τοπομπία ("in order that the people may always have their own proper maritime commerce and supply of grain");[36] the Athenian decree of ca. 330 B.C., *IG* II^2 416, lines 10–11, ὅπως ἂν σῖτος ὡς ἀφ[θονωτάτως]| εἰσπλεῖ τῶι δήμωι τ[ῶι Ἀθηναίω]|ν ("in order that as plentiful a supply of grain as possible may sail in for the Athenian people"); Migeotte 1992, no. 74, lines 4–5, Iasos, 2nd century B.C., [ὅπως] ὁ δῆμος ἀεὶ εὐ[δαιμόνως]| [ζῆι ἐν δαψ]ιλείαι σίτου γιν[όμενος] ("in order that the people may always live prosperously, enjoying a plentiful supply of grain").

[35] For this information I am grateful to J. McK. Camp II and M. B. Richardson, who are preparing the editio princeps.

[36] For the correct accentuation of the two nominative nouns and the interpretation of lines 217–232, see Bresson 1993, pp. 171–177.

Elsewhere in our law ὁ δῆμος is explicitly the *ekklesia*, lines 36, 42, 44, 49, 53, and 54. In the opening purpose clause, however, the word has a wider meaning. It is for the Athenian *people* as a whole that the lawgiver is making grain available. For helpful discussion of this distinction, see Hansen 1978a, pages 128–141.

ἐν τῶι κοινῶι is here implicitly contrasted, as so often, with the private sphere, which was the normal source of grain for Athenian citizens. Although the Athenians regulated the importation of grain into their city by means of both laws and administrative officials, the trade itself was almost entirely in the hands of private shipowners, merchants, and retailers who sold to individuals mainly on their own discretion. For a convenient list of Athenian regulations on the grain trade, see Isager and Hansen 1975, pages 28–29; Gauthier 1981. See below, page 68.

Agyrrhios' law now mandates more direct state intervention in the transportation, storage, handling, and sale of a certain portion of imported wheat and barley at Athens. The grain regulated by his law belongs to the *demos*, for it originates as a state-imposed tax in the Athenian-held islands of Lemnos, Imbros, and Skyros. It is this grain that the Athenians, through their elected officials, will now weigh, store, and sell to the *demos*, ἐν τῶι κοινῶι, as opposed to the grain for sale on the open market in the hands of private merchants. The latter grain, as Gauthier (1981, pages 17–19) observed, was called ὁ ἐν ἀγορᾷ σῖτος. It is this contrast that my translation, "in the public domain," is meant to convey.[37]

But τὸ κοινόν, as Schmitt-Pantel (1992, pages 107–113, 478–479) has ably demonstrated, has a wider resonance, one that is, in my view, pertinent to the purpose of Agyrrhios' law. As she observes, "Tout *koinon* contient donc en lui la virtualité du partage" (page 109). Like the sacrificial meat distributed by the *polis* at the great Athenian festivals, the grain of the *demos* here is something in which all, or most, Athenians will have a share.

The concrete sense of τὸ κοινόν, however, as a repository for public resources may also be present here, for, as we shall see, the *polis* will provide a large and convenient public storehouse for its grain. Our law thus contains new evidence for a Greek community of the Classical period establishing a public facility for bulk storage of public grain. We ought, therefore, to qualify notions like those of Gallant (1991, pages 179–181) to the effect that Greek *poleis* did not do this, because they "were neither strong enough nor bureaucratized enough to extract sizeable quantities of food from their citizenry" and that "there were structural impediments to the implementation of strategies based on mass storage at the communal level."[38]

[37] For this less concrete sense of τὸ κοινόν, cf. Herodotus 5.85, πεμφθέντες ἀπὸ τοῦ κοινοῦ ("dispatched from the *koinon*"); 8.135.2; Thucydides 1.89.3, with the note of Hornblower (1991, pp. 135–136); 2.37.2; Plato, *Laws* 3.681C; Plutarch, *Moralia* 644C; *IG* I^3 40, line 11; 101, line 53; 118, lines 12–16 (cf. *SEG* XXXI 19); and the common phrase κοινῆι καὶ ἰδίαι, *IG* II^2 252, lines 5–6; 339 b, lines 4–5; 398 b, lines 3–4; 399, line 13; cf. Isokrates 7.52; Xenophon, *Poroi* 4.33, ὡς ἂν ἡγοῦμαι κατασκευασθείσης τῆς πόλεως ἱκανὴν ἂν πᾶσιν Ἀθηναίοις τροφὴν ἀπὸ κοινοῦ γενέσθαι ("how I believe that the *polis* should be organized so as to ensure an adequate level of public maintenance for all Athenians"); and the discussions by Charneux (1984, pp. 210–211), with earlier bibliography; Musti (1985); and Tréheux (1987, pp. 38–46). For a good example of "public grain," see the decree from Koroneia in Boiotia, *SEG* XLIII 205, lines 23–24, καθὼς κὴ τὸν λυπὸν σῖτον τὸν δαμόσιον with discussion in Migeotte 1993; forthcoming, "Ventes."

[38] For the concrete sense of τὸ κοινόν as the state treasury, see Herodotus 7.144.1; 9.87.2; Thucydides 1.80.4; 6.6.3; *IG* I^3 245, line 10 (deme). For τὰ κοινά as χωρία, see the deme decree of Teithras, ca. 350 B.C., *SEG* XXIV 151, lines 2–5, ὅπως ἂν σᾶ ἦι τ|οῖς δημόταις τὰ κοινὰ καὶ εἰδῶσι Τειθράσιοι τὰ ὑπάρχ|ο[ντα] καὶ τὰ προσιόντα, ἀναγρά[ψαι τὸ]ν δήμαρχο|[ν] ὁπ[όσ]οι κατάπαξ μεμίσθωνται τῶν κοινῶν ("in order to safeguard the *koina* of the demesmen and so that the Teithrasians will know the extent of their holdings and their income, let the *demarchos* write up every single one of those who have leased their common property").

LINES 6–8

τὴν δωδεκάτην πωλ[εῖ]-
ν τὴν ἐν Λήμνωι καὶ Ἴμβρωι καὶ Σκύρω[ι κ]-
αὶ τὴν πεντηκοστὴν σίτο

Sell the $8\frac{1}{3}$% tax [which originates] in Lemnos, Imbros, and Skyros and the 2% tax in terms of grain

To my knowledge, the present law is our first and only evidence for an Athenian tax at the rate of $8\frac{1}{3}$%, ἡ δωδεκάτη.[39] We hear elsewhere of Athenian taxes levied in the following denominations: $\frac{1}{5}$%, πεντακοσιοστή;[40] 1%, ἑκατοστή;[41] 2%, πεντηκοστή;[42] 2.5%, τετταρακοστή;[43] 5%, εἰκοστή;[44] 10%, δεκάτη.[45] With the exception of the εἰκοστή of the Peisistratidai, which seems to have been a tax in kind, most scholars have concluded that all these other Athenian taxes, including the *eisphora*, were collected in cash.[46] But there is much about the entire system of Athenian taxation that remains obscure. Apart from the *eisphora*, the whole topic would repay closer scrutiny.[47] We

[39] In 14.27, Demosthenes hypothetically proposes this rate of taxation for an *eisphora*, only to dismiss it as something that the Athenians would not endure. For a tax of $8\frac{1}{3}$% on wine in Iasos, ca. 200 B.C., see *SEG* XLI 929, lines 4–5, πλὴν τοῦ περιγινομένου| [ἀπὸ] τῆς δωδεκάτης τοῦ πωλ[ουμένου ο]ἴνου τοῦ ἐπιχωρίου ("except for the sum left over from the $8\frac{1}{3}$% tax on the sale of local wine"). Elsewhere the rate of $8\frac{1}{3}$% appears as the upper limit for interest at Ephesos, ca. 297/6 B.C., τόκους δὲ αὐτοῖς εἶναι μὴ πλείους δωδεκάτων ("their rate of interest is not to exceed $8\frac{1}{3}$%"), *Syll.*³ 364, line 74 (*I. Ephesos* Ia, 4). For 8% as a common rate for rental of land in Athens, see below, p. 32.

[40] Aristophanes, *Ekklesiazousai* 1006–1007.

[41] [Xenophon], *Ath* 17; Aristophanes, *Wasps* 658; Demosthenes 14.27; *IG* II² 1580, 1594–1603, the *rationes centesimarum*, on which see Rosivach 1992 and Lambert 1997.

[42] Andokides 1.133–134 (above, pp. 19–20); Euboulos, *PCG* V, F 119 (= Athenaios 2.49c); Demosthenes 14.27; 24.120; [Demosthenes] 34.7; 35.29–30; 59.27; Hypereides F 106 (= Harpokration, *s.v.* πεντηκοστή); Lykourgos, *Leokr.* 19; Menander, F 897 Koerte (= Photios, *s.v.* πεντηκοστολόγοι); Pollux 8.132; 9.28–29; *Et. Mag.* *s.v.* πεντηκοστευόμενον; *Souda Lex.*, *s.vv.* πεντηκοστή, πεντηκοστολόγιον; Bekker, *Anecdota Graeca* 1.192.30; 1.297.21–26; *IG* I³ 133, line 25, see below, pp. 37–38, note 78; *IG* II² 334 + *SEG* XVIII 13, lines 11–13; *IG* II² 1635, line 38 (*I. Délos* 98). Omitted from this list is Demosthenes 21.133, for the *pentekostologoi* there mentioned are probably Chalkidian, not Athenian; see Knoepfler 1981, pp. 328–329; *contra* MacDowell 1990, p. 353. For an excellent illustration of how the *pentekoste* worked in Kyparissia, see *IG* V.1.1421. On the *pentekoste* in general, see Gofas 1993, pp. 225–227.

[43] Aristophanes, *Ekklesiazousai* 823–825, with the scholia. Thomsen (1964, pp. 184–185) persuasively argued that these first two taxes, and the ἑκατοστή, were indirect taxes having no connection with the *eisphora*.

[44] Thucydides 6.54.5; 7.28.4; Aristophanes, *Frogs* 363; Pollux 9.29; Bekker, *Anecdota Graeca* 1.185.21. *IG* II² 24, lines 3–5; 28, lines 7–8. For Thucydides 6.54 on the *eikoste* tax of the Peisistratidai, cf. Aristotle, *Ath* 16.4, who makes it a *dekate*, Rhodes 1981, p. 215; Andrewes, *CAH*² III.3, pp. 407–408; Millett 1991, pp. 263–264, note 44. On the *eikoste* elsewhere, see Gauthier 1989, pp. 33–36.

[45] E.g., Xenophon, *HG* 1.1.22; 4.8.27, 31; Demosthenes 20.60; 22.77; 24.120; Aristotle, *Ath* 16.4 (above, note 44); Polybios 4.44; Pollux 8.132; 9.28–29; Harpokration, *s.vv.* δεκατεύειν, δεκατευτάς, δεκατηλόγους; Bekker, *Anecdota Graeca* 1.185.21; *IG* I³ 52, line 7 (see below, pp. 82–84); *IG* II² 1609, col. II, line 97.

[46] For possible *eisphora* payments in kind, see the bibliography in *SEG* XXXIX 172. For Athenian taxes expressed in specific sums, not as percentages, see the five-drachma tax for Theseus (lines 479–480) and the one-drachma tax for Asklepios (lines 487–488) in *IG* II² 1582 + (*Agora* XIX, P26); the one-drachma tax for Apollo Delios in *IG* I³ 130, lines 6–7 (cf. Lewis 1960); the annual tax of two drachmai for the *hippeis*, one drachma for the hoplites, and three obols for the *toxotai* for Apollo in *IG* I³ 138, lines 1–4. On these taxes, see Schlaifer 1940, pp. 233–241; Parker 1996, p. 125.

[47] Boeckh (1886, pp. 366–368), of course, made a magnificent start. Standard references to Athenian taxation include W. Schwahn, *RE* XXXIII (1936), cols. 843–845, *s.v.* Νόμος τελωνικός; *ibid.*, V.A (1934), cols. 226–262, *s.v.* Tele; 418–425, *s.v.* Τελῶναι; Andreades 1933, pp. 125–161, 277–285, 294–299, 326–349; Heichelheim 1964, II.133–148, 238–245; Thomsen 1964; Jones 1974; Gera 1975; Velissaropoulos 1980, pp. 205–210; Brun 1983; Littman 1988; Isager and Skydsgaard 1992, pp. 135–144.

It is a pity that D. M. Lewis' call (1959, p. 243) for a thorough re-examination of the entire Athenian system of taxation has not yet been answered. It has been recently echoed by Pritchett (1991, p. 482). The racy formulation of De Ste. Croix (1981, p. 206), "It is a melancholy fact, characteristic of our sources of information for Greek—even Athenian—economic history, that our fullest list of taxes for a single city in any literary source should occur in Comedy: Aristophanes, *Wasps* 656–60!", should not obscure the fact that the time is ripe to harvest the rich evidence on taxation that has accumulated in Attic inscriptions published since the days of Boeckh and Andreades. See the timely observations of Migeotte

will see that our new law permits the conjecture that taxes in kind might not have been as unusual at Athens as is sometimes assumed.

After stating that the purpose of his law is to provide the demos with grain, not money, Agyrrhios' first instruction is to "sell the $8\frac{1}{3}$% tax (*dodekate*)," τὴν δωδεκάτην πωλ[εῖ]ν, and he defines this tax precisely as originating on the three islands, τὴν ἐν Λήμνωι καὶ Ἴμβρωι καὶ Σκύρω[ι]. The verb πωλέω appears again in line 41, this time in the imperative, πωλόντων, with the ten men (δέκα ἄνδρας, lines 36–37, οὗτοι, line 40) as its subject and with grain (τὸν σῖτον, line 40) as its object. Again in lines 42 and 46 the infinitive πωλε̃ν twice designates the same activity of the ten men, who are clearly selling the wheat and barley, τὴν τ[ι]μὴν τῶν πυρῶν καὶ τῶν κριθῶν ὁπόσου χ[ρ]ὴ πωλε̃ν τοὺς αἱρεθέντας, lines 44–46.[48] This activity of the ten men is described again in lines 51–52 as ἐπειδὰν δὲ ἀποδῶνται οἱ αἱρεθέντες τὸν σῖτον ("whenever the elected men sell the grain)." Throughout this passage in lines 41–52 the verb πωλέω with τὸν σῖτον as its expressed or understood object clearly means "to sell" the grain. It is this selling, to be examined in detail later, that puts *sitos* into the hands of individual members of the *demos*. The selling of the grain to the *demos* in the Agora by the ten men is, however, only the last step in an elaborate process that begins in line 5.

Here, at the beginning of the law, the verb πωλέω takes as its object, not τὸν σῖτον, but τὴν δωδεκάτην, the $8\frac{1}{3}$% tax, and τὴν πεντηκοστήν, the 2% tax. To sell (πωλεῖν) or to buy (ὠνεῖσθαι) a tax in ancient Athens is to sell or buy the right to collect such tax, cf. Aristotle, *Ath* 47, οἱ πωληταὶ . . . τὰ μέταλλα πωλοῦσι καὶ τὰ τέλη ("the *poletai* sell the [contracts] for the mines and [the rights to collect] taxes"); *IG* II^2 334 + *SEG* XVIII 13, lines 11–12, [τοὺς δὲ πωλητὰς τὴ]ν πεντηκοστὴν πωλεῖν τὴν ἐν τῆ[ι Νέαι] ("let the *poletai* sell the [right to collect] the 2% tax in the Nea"); Aischines 1.119, ἡ βουλὴ πωλεῖ τὸ πορνικὸν τέλος ("the council sells [the right to collect] the tax on prostitutes"). The meaning of τὴν δωδεκάτην πωλ[εῖ]ν in line 6 is different from τὸν σῖτον πωλε̃ν in lines 41–52. There is a similar shift in the objects governed by this verb in Aristotle, *Ath* 47, where the *poletai* both πωλοῦσι καὶ τὰ τέλη ("sell the right to collect the taxes") and τὰς οὐσίας . . . πωλοῦσιν ("sell the [confiscated] properties"). See the excellent detailed discussion of this verb by Chantraine (1940), and the good notes in Rhodes 1981, pages 552 and 554. The *polis* or its representatives, who are the unexpressed subject of the infinitive πωλ[εῖ]ν in lines 6–7, are not instructed to sell grain or any other commodity to the "purchasers" of the *dodekate* and the *pentekoste*. They sell them the right to collect these taxes in Lemnos, Imbros, and Skyros. See further below, pages 51–52.

Although the infinitive πωλ[εῖ]ν in line 6 introduces a long and complex section containing many details as to the conditions under which the *dodekate* and *pentekoste* taxes are to be "sold," i.e., leased or farmed out, the law does not specify the public official(s) who are to conduct the "sale." Nor is the venue of the sale mentioned here. We are also left to infer from later parts of the law the nature of the procedure through which the *polis* will choose "buyers" of the right to collect the $8\frac{1}{3}$% and the 2% tax in the three islands. It is likely that Agyrrhios omitted explicit mention of these items here because the Athenians had already formulated normal procedures for farming out taxes in existing regulations such as οἱ νόμοι οἱ τελωνικοί, mentioned by Demosthenes (24.96–98, 101, and 122). Building on this foundation, Agyrrhios may have spelled out in his new law only those provisions that departed from or supplemented current practice. We shall see that

(1995, p. 10), "Mais la masse de l'information vient de l'épigraphie. Cette documentation, qui ne cesse de s'enrichir, est loin d'avoir été exploitée comme elle le mérite, sans doute parce qu'elle se présente de manière dispersée et rebutante pour les profanes. Un travail considérable reste donc à faire, dont les résultats permettront de combler des lacunes, de revoir des doctrines établies—par exemple celle concernant les taxes et les impôts de type foncier, qui étaient sans doute plus répandus qu'on ne l'admet généralement." For the little that is known about deme taxation, see Whitehead 1986, pp. 150–152. For the even more meager evidence for taxation by phratries, see Lambert 1993, pp. 304–305.

[48] I do not believe that the variation in spelling between πωλ[εῖ]ν, lines 6–7, and πωλε̃ν, lines 42 and 46, is significant; cf. σίτου, line 3, σίτο, lines 8, 39, 55.

such provisions were both numerous and complex. It is only natural that the lawgiver devoted the bulk of his new law to them. Helpful on the νόμοι τελωνικοί at Athens are W. Schwahn, *RE* XXXIII (1936), cols. 843–845, *s.v.* Νόμος τελωνικός; Gofas 1969, page 370, note 1; Migeotte, forthcoming, "Aspects."

It is possible to suggest a plausible venue for the procedure of "selling" the *dodekate* from the fact that it is the *boule* that is required to approve two guarantors for each portion (μερίς) of wheat and barley that a "buyer" of the right to collect the *dodekate* and the *pentekoste* undertook to deliver to Athens (lines 29–31), ἐγγυητ⟨ὰ⟩ς . . . οὓς ἂν ἡ βουλὴ δοκιμάσηι. We have seen from Andokides 1.134, quoted above, page 19, that it was in the *boule* that Andokides claimed to have outbid Agyrrhios and his associates in 401/0 B.C. when the right to collect the *pentekoste* was being sold by auction, παρελθὼν ἐγὼ εἰς τὴν βουλὴν ὑπερέβαλλον ἕως ἐπριάμην ἓξ καὶ τριάκοντα ταλάντων ("coming into the council, I outbid them until I bought [the right to collect the tax] for 36 talents"). As we have seen, Aischines (1.119) observed that the *boule* annually "sold" the tax on prostitutes; quoted above, page 28. It is probably in the *boule* that another tax, the *pentekoste* in the Nea, is farmed out in the law of the Lesser Panathenaia of ca. 336–334 B.C., *IG* II^2 334 + *SEG* XVIII 13, lines 11–15.

τοὺς δ]-
[ὲ πωλητὰς τὴ]ν πεντηκοστὴν πωλεῖν τὴν ἐν τῆ[ι Νέαι χω]-
[ρὶς τῶν ἄλλω]ν· τοὺς δὲ πρυτάνεις προγράφει[ν βουλῆς]
[ἕδραν εἰς τὴ]ν μίσθωσιν τῆς Νέας διαρρήδην [καὶ τῆς π]-
[εντηκοστῆς] τὴν πρᾶσιν τῆς ἐν τῆι Νέαι χωρὶ[ς . . .6. . .].

Let the *poletai* sell [the right to collect] the 2% tax in the Nea apart from the others. Let the *prytaneis* give written notice in advance of a session of the council for the express purpose of leasing the Nea and for the sale of [the right to collect] the 2% tax in the Nea separately.

Participation of the *boule* in farming out taxes is also attested in Aristotle, *Ath* 47: οἱ πωληταὶ ί μέν εἰσι, κληροῦται δ' εἷς ἐκ τῆς φ[υ]λ[ῆς. μ]ισθοῦσι δὲ τὰ μισθώματα πάντα, καὶ τὰ μέταλλα πωλοῦσι καὶ τὰ τέλη μ[ετὰ τ]οῦ ταμίου τῶν στρατιωτικῶν καὶ τῶν ἐπὶ τὸ θεωρικὸν ῃρημένων ἐναντίον τῆς [βουλῆς] . . . καὶ τὰ τέλη τὰ εἰς ἐνιαυτ[ὸ]ν πεπραμένα ἀναγράψαντες εἰς λελευκωμένα γραμματεῖα τόν τε πριάμεν[ο]ν καὶ ὅσα ἂν πρίηται τῇ βουλῇ παραδιδόασιν. ἀναγράφουσιν δὲ χωρὶς μὲν οὓς δεῖ κατὰ πρ[υτ]ανείαν ἑκάστην καταβάλλειν εἰς δέκα γραμματεῖα, χωρὶς δὲ οὓς τρὶς τοῦ [ἐ]νιαυτοῦ, γραμματεῖον κατὰ τὴν καταβολὴν ἑκάστην ποιήσαντες, χωρὶς δὲ οὓς [ἐπ]ὶ τῆς ἐνάτης πρυτανείας ("Next there are the ten *poletai*, one chosen by lot from each tribe. They let out all the [public] contracts and they offer for lease the mines and the taxes in the presence of the *boule*, together with the treasurer of the military fund and those elected to manage the theoric fund . . . and after recording on whitened tablets the taxes that have been sold for the current year and the purchaser and the purchase price, they turn these over to the *boule*. They record separately on ten tablets those who must make a payment each prytany, and those who must make a payment three times in the year they record on a separate tablet for each installment, and separately those who must pay in the ninth *prytany*"). See Rhodes 1981, *ad loc.*, and for the role of the *boule*, Rhodes 1972, pages 96–98.

I conclude from this evidence that the selling of the *dodekate* tax on the grain in the islands of Lemnos, Imbros, and Skyros and the *pentekoste* probably took place in the *boule*.

Aristotle, *Ath* 47, demonstrates that in the early 320s the *boule* was assisted in farming out taxes by the *poletai*, the *tamias* of the stratiotic fund, and the controllers of the theoric fund. Agyrrhios' law mentions none of these officials, although as the final step in the entire process outlined on our stone, the proceeds of the sale of the grain from the islands went to the stratiotic fund, lines 54–55. Its treasurer may have been involved at this point, as we shall see below, page 77. Presumably, when the *boule* had sold the right to collect the *dodekate* and the *pentekoste*, someone had to record the names of the "purchasers" and the amounts of grain they contracted to deliver, either singly or as participants in a *symmoria*. It was also necessary to record the names of their guarantors and to verify payment of the ἐπώνια and κηρύκεια (line 28). Perhaps the most logical candidates for these duties are the ten *poletai* who performed such tasks later in Aristotle's day. D. M. Lewis has plausibly restored the *poletai* as farming out a tax, possibly connected with the shrine of Apollo Delios at Phaleron, ca. 432 B.C., *IG* I^3 130, lines 6–7.[49] For a possible association of the *poletai* with another tax connected with the sanctuary of Bendis in the Peiraieus, see *IG* I^3 136, lines 22–23, with the important discussion by Ferguson (1949, pages 142–143). The *poletai* were certainly in active participation with the *boule* in 375/4 B.C., as is clear from Nikophon's law on silver coinage, *SEG* XXVI 72, lines 47–49.[50] It is reasonable to suggest that they assisted the *boule* in selling the *dodekate* and the *pentekoste* in the next year and that they therefore qualify as one of the unexpressed subjects of the infinitive πωλ[εῖ]ν in lines 6–7. Langdon (1994, pages 258–261) is helpful on the interaction of the *poletai* and the *boule*.

Finally, modern readers of his text might have been better informed about the procedure of "selling" the *dodekate* if the lawgiver had spelled it out in lines 8 and following, instead of plunging directly into a detailed definition of a μερίς, lines 8–10, and the duties of "the purchaser," lines 10–15. Again, I suggest here that the procedure for "selling a tax" had already been described in other legislation, thus sparing Agyrrhios a laborious recital of normal administrative practice here. Note that in the parallel instructions for "selling" the *pentekoste* tax in the Nea in *IG* II^2 334 + *SEG* XVIII 13, lines 11–15, quoted above, page 29, few details are given about procedure. The *poletai* are simply told to sell the tax—[τοὺς δὲ πωλητὰς τὴ]ν πεντηκοστὴν πωλεῖν τὴν ἐν τῆ[ι Νέαι]—and everyone knew what that meant.

We, however, have to reconstruct the procedure of the "selling" of the *dodekate* and the *pentekoste* on the basis of inferences drawn from the text of the law itself and from parallel passages that describe the "sale" of other taxes. In the law itself perhaps the most useful clues are provided in lines 27–31 where Agyrrhios states, as we shall see, that the purchaser of the *dodekate* will not be required to make a down payment (προκαταβολή), but he will be charged on each μερίς a sum of twenty drachmai that will cover ἐπώνια καὶ κηρύκεια. These two expenses are attested elsewhere in procedures for the leasing of property and the letting of contracts by the *poletai*. We examine their meaning in more detail in the commentary on line 28. Now it is enough to note that κηρύκεια are the fees paid to a herald who served as auctioneer. It is a valid inference, therefore, from the fact that the purchaser of these two taxes had to pay an auctioneer's fee, that the process of selling the *dodekate* and the *pentekoste* included an auction. This should not surprise us, for we have seen that it was in an auction held before the *boule* that Andokides claimed to have outbid Agyrrhios and his associates for the right to collect the *pentekoste* tax, 1.133–134, above, page 19. A public auction at Athens for the sale of state taxes is also the setting for an anecdote about Alkibiades and a metic related in some detail by Plutarch (*Alkibiades* 5). The Athenian general urges the metic to τοὺς ὠνουμένους τὰ τέλη τὰ δημόσια ταῖς τιμαῖς ὑπερβάλλειν ἀντωνούμενον ("to oppose those who are trying to buy [the right to collect] the public taxes by bidding up the price"), which he then does by a margin of a talent. Alkibiades confounds the competing τελῶναι by standing in as a guarantor, ἐγγυητής, for the metic, thus forcing them to bribe the metic to withdraw from the auction by paying him a talent.[51] Since, then, the selling of a tax by the *poletai* and/or the *boule* seems normally

[49] Ἀποδιδόσθον οἱ π[ολεταὶ τὸ τε͂ς δραχμε͂ς τέλος– – –] ὅταμπερ τὸ το͂ν ἀπα[ρχο͂ν τέλος μισθο͂σι] ("let the *poletai* sell [the right to collect] the one-drachma tax . . . whenever they lease out the tax on the first-fruits"). I would prefer to restore the last word as πολο͂σι.

[50] See Langdon's good discussion of this passage in *Agora* XIX, p. 63, and his collection of epigraphical evidence for *poletai* selling the privilege of collecting state taxes on p. 65.

[51] The point of this story is lost if Plutarch fails to understand the mechanics of selling the right to collect public taxes at auction. Since his information squares with all other ancient evidence for this procedure, I cannot agree with Perrin's rejection (1912, p. 264) of this valuable

to have been conducted by means of an auction,[52] I suggest that the *dodekate* and the *pentekoste* in Lemnos, Imbros, and Skyros were also sold in this manner.

Although in the heading in lines 3–4 the law is labeled νόμος περὶ τῆς δωδεκάτης τοῦ σίτου τῶν νήσων, "law concerning the *dodekate* on the grain of/from the islands," in the text itself Agyrrhios seems to have taken care to define the tax specifically as τὴν δωδεκάτην πωλ[εῖ]ν τὴν ἐν Λήμνωι καὶ Ἴμβρωι καὶ Σκύρωι (lines 6–7). To my knowledge, only two other Athenian taxes are similarly designated with a locative preposition. The first is the δεκάτη, or 10% tax, that Athens imposed on the cargoes of ships sailing out of the Black Sea. In 410 B.C. the Athenians built a δεκατευτήριον at Chrysopolis on the Hellespont in order to collect this tax on the spot. Xenophon describes this tax in 389 B.C. as ἡ δεκάτη τε τῶν ἐκ τοῦ Πόντου πεπραμένη εἴη ἐν Βυζαντίωι ὑπ' Ἀθηναίων ("the 10% tax on goods leaving the Pontos was sold in Byzantion by the Athenians"), *HG* 4.8.31.[53] The second example is the *pentekoste* in the Nea, which we have already had occasion to note in the law on the Lesser Panathenaia, *IG* II^2 334 + *SEG* XVIII 13, lines 11–12, [τὴ]ν πεντηκοστὴν πωλεῖν τὴν ἐν τῆ[ι Νέαι]. Robert has taught us that this wording is precise. The *pentekoste* was a 2% tax that was farmed in Athens but originated, i.e., was levied and collected, "in the Nea."[54] If these two parallels can be pressed, we can probably conclude that the *dodekate* on the grain from Lemnos, Imbros, and Skyros was a tax imposed upon and collected locally in the islands themselves.[55]

In 374/3 B.C. the Athenians were in a position to pass a law imposing or redefining a tax on the wheat and barley grown in these three islands because each was at that time an Athenian possession. The inhabitants seem for the most part to have been Athenian citizens.[56] The islands in a sense "belonged" to the Athenians.[57] More than a decade earlier, in 387/6 B.C., the Athenian council and assembly passed and had inscribed on a stele on the Acropolis a lengthy and detailed decree regulating land tenure, residency, and probably other requirements for the inhabitants and klerouchoi on Lemnos.[58] Similar regulations may have been published for Skyros and Imbros at the same time. It is not difficult to imagine that Athenian legislation governing their own citizens on these islands had included provisions on taxation.[59] In 374/3 B.C. the Athenians by law imposed

passage, "The anecdote cannot be regarded as a source for exact economic detail."

[52] For an excellent, concise statement of normal procedure, see Youtie 1967, pp. 8–10; cf. Langdon, *Agora* XIX, pp. 62–65. Hallof (1990) and Langdon (1994) examine the involvement of the *poletai* in such auctions at Athens.

[53] Cf. Xenophon, *HG* 1.1.22; 4.8.27; Polybios 4.44.4; Diodoros 13.64.2. On this tax or toll, see below, pp. 82–84.

[54] *Hellenica* XI–XII, 1960, p. 193. Robert interpreted the tax as applying to imports and exports at the harbor of coastal Oropos. While accepting this view of the nature of the tax, Langdon (1987, pp. 55–58) identified Nea as an island to the east of Lemnos, now submerged. In re-editing the law in *Agora* XIX, pp. 184–186, L7, Walbank adopts the position of Robert. Gauthier (*BE* [1988] 349) rejects Langdon's view, stressing the consistent use of the article, ἡ Νέα, throughout the text.

[55] For an example of Athenians imposing a tax in Potideia at the *situs* instead of at the place of domicile, see [Aristotle], *Oikonomika* 2.1347a, with the good discussions of Bullock (1939, pp. 128–129) and Salomon (1997, pp. 202–208). This procedure could also be relevant in the case of Athenians holding *kleroi* on Lemnos, Imbros, and Skyros. See Tréheux 1991, pp. 143–147.

[56] For a helpful review of the evidence for the Athenian citizenship of the inhabitants, see Gschnitzer 1958, pp. 101–103; Graham 1983, pp. 167–192; Schmitz 1988, pp. 79–115; Figueira 1991, pp. 66–73; Cargill 1995, pp. 1–8, 12–15, 42–66, 92–109; Salomon 1997, pp. 76–81, 91–95.

[57] Andokides 3.12, νῦν δὲ ἡμετέρας εἶναι ('now they are ours"); Xenophon, *HG* 5.1.31, ταύτας δὲ ὥσπερ τὸ ἀρχαῖον εἶναι Ἀθηναίων ("these [islands] are to belong to the Athenians just as they did originally"); Diodoros 16.21.2, Ἴμβρον μὲν καὶ Λῆμνον οὔσας Ἀθηναίων ("Imbros and Lemnos belong to the Athenians"). This point is well brought out by Gschnitzer (1958, pp. 99–100), "das Gebiet der attischen Aussensiedlungen ist athenisches Territorium."

[58] *IG* II^2 30, as supplemented and re-edited by Stroud (1971, pp. 162–173, no. 23); reprinted by Walbank in *Agora* XIX, L3 and by Woodhead in *Agora* XVI, no. 41. The length (at least 53 lines of ca. 95–100 letters each) and complexity of this inscription have not always been appreciated by students of Athenian klerouchies; conspicuous exceptions are Bugh 1988, pp. 209–218 and Salomon 1997, pp. 142–150. See also *SEG* XLII 87.

[59] In fact taxes seem to be mentioned in an uncertain context in line 10 of the decree on Lemnos cited

a produce tax on these three islands in such a way as to produce for the *demos* at home wheat and barley that was to be brought to the Sanctuary of Aiakos. Our law says nothing about how this tax was to be collected on the islands, nor do we know if it applied also to grain produced on land not held by Athenians—if indeed any such land existed.[60] It is even possible that the grain-tax was viewed in Athens as the equivalent of a fee or a kind of rent charged to the inhabitants and klerouchoi for the privilege of holding land on Lemnos, Imbros, and Skyros,[61] but our law is silent on such matters. We do know, however, of rents in kind at this period, and 8% is (suggestively?) a very common rate for rent paid by lessors of public land in Attica; see Osborne 1988, pages 285, 323; Walbank 1983, page 217. In 329/8 B.C. the famous orator Hypereides paid his rent for farmland he leased in the Rarian Plain in kind, in barley, *IG* II^2 1672, lines 252–262. We do not know the rate at which he was charged, but over a four-year period he paid a total of 2,732 *medimnoi*. For rents in kind, see also, e.g., *SEG* XXXVIII 380. For possible repayment of a loan in kind (σῖτος) relating to land, see Lysias 32.15; Cohen 1989, page 210, note 16.

Restored officially to Athens under the terms of the King's Peace,[62] not only were these three islands strategically located on the route of the grain ships sailing from Chalkidike, Thrace, and especially from the Hellespont to the Peiraieus,[63] but each also produced significant quantities of wheat and barley of its own.[64] In his *First Philippic* of ca. 351 B.C. Demosthenes claims that σῖτος is plentiful on Lemnos in winter (4.32). Our best ancient evidence for grain production on Lemnos, Imbros, and Skyros consists of a frequently discussed passage in the inscribed accounts of the Epistatai of Eleusis and the Treasurers of the Two Goddesses for 329/8 B.C., *IG* II^2 1672. Beginning in line 263 are recorded τῆς ἐπαρχῆς/ἀπαρχῆς[65] τοῖν θεοῖν τοῦ σίτου

in note 58. Salomon (1997, p. 183) suggests that the grain tax of $8\frac{1}{3}$% may have been introduced in Lemnos shortly after 387/6 B.C. See below, p. 109. For earlier examples of the Athenian Assembly imposing an Athenian tax on Athenian settlers in Salamis, see *IG* I^3 1, line 3; in Chalkis, *IG* I^3 40, lines 52–57 (doubtful; see Lewis' good note *ad loc.*); and Hestiaia, *IG* I^3 41, lines 36–38, with the persuasive analysis of Graham (1983, pp. 170–172). For a brief discussion of the limited extent of "ein autonomes Finanzwesen" in Athenian klerouchies, based mainly upon [Aristotle], *Oikonomika* 2.1347a (on which see above, note 55), see Gschnitzer 1958, pp. 109–110.

[60] In *IG* XII.8.4 (undated) there is evidence that "the demos of Athenians living in Myrina" on Lemnos granted a χωρίον to "the Chalkidians living in Myrina." Some of these Chalkidians may have been given land to farm. Fredrich *ad loc.* would date this ca. 348 B.C. Cargill (1995, pp. 66–77) has a helpful discussion and bibliography on this inscription and in general on non-Athenians on the three islands at this time. See also Salomon 1997, p. 183, note 598.

[61] This is a delicate point on which perception may have been more important than fact or definition. Gernet (1909, pp. 352–353) argued that "nous ne voyons à aucun moment que les clérouques doivent payer d'un fermage quelconque la propriété que leur a concédée la cité ... sans réglementation légale, sans appareil administratif, les clérouchies fournissaient aux besoins d'Athènes, soit en alimentant le commerce, soit par des envois directs et gratuits. ... C' est un régime d'individualisme instinctif et souple." See also Garnsey 1988, p. 131; Isager and Skydsgaard 1992, p. 140–141; Salomon 1997, p. 183, note 598: "Non crea problema l'imposizione di una tassa in natura a cittadini ateniesi, se si considerano Lemno, Imbro e Sciro come bacini da cui Atene può tranquillamente attingere per approvvigionarsi."

[62] Andokides 3.12; Xenophon, *HG* 5.1.31. For the status of Lemnos prior to 387/6 B.C., see Gschnitzer 1958, pp. 100, 110–112; Stroud 1971, pp. 170–173; Graham 1983, pp. 175–190; Figueira 1991, pp. 253–256; Cargill 1995, pp. 59–66; Salomon 1997, pp. 31–66.

[63] For a timely reminder of the role of Athenian klerouchoi in both the defense of Lemnos and in keeping the peace between the *demoi* of Myrina and Hephaistaia, see Salomon 1997, pp. 120–139, 184–188. For grain ships sailing to Lemnos from the Hellespont, see Demosthenes 18.77–78, with Gernet 1909, p. 362.

[64] For testimonia on these regions as important sources of grain for Athens, with helpful discussion, see Gernet 1909, pp. 314–326, 350–353. For a coin of Imbros with an ear of grain as a symbol, see Imhoof-Blumer 1882, p. 147, no. 3. *IG* XII.8.51, line 19, of the 2nd century B.C. lists 530 *medimnoi* of wheat stored in a sanctuary (of Athena?) on Imbros. Aristotle, *Politics* 1.1259a1, mentions Apollodoros of Lemnos, a writer on agriculture; see Cargill 1995, p. 270, no. 126. For the value of these three islands to Athens both as centers of production of grain and as strategic stations for the grain ships, see Cargill 1995, pp. 4–6, 12–15, 42–58; Salomon 1997, pp. 175–188.

[65] These two words are apparently used interchangeably in this inscription; see lines 182, 288, and 297 with Kirchner's note on 297.

κεφάλαια ("totals of the first-fruits of grain for the two goddesses"). The Epistatai first list the totals of wheat and barley sent as first-fruits to the two goddesses from Attica. The amounts of grain, expressed in *medimnoi* and fractions thereof, are arranged by the ten Attic tribes in their regular order. There follow amounts of wheat and barley brought from Drymos, from τῆς ἐπ' Ἀμφιαράου,[66] and from Salamis. Then we read in lines 275–279, ἐΣκύρου στατη[γ]ὸς [*sic*] Μνησίστρατος Κυθήρ κρι ΔΔΔΔΠỊỊỊ, πυρῶν ὀκτώ·| ἐγ Μυρίνης στατηγὸς [*sic*] Σθένυλλος Εἰρεσίδης, Σώπ̣[ατρο]ς Ἀλωπεκῆθεν, Αἰσχύλος Ἐρχι[ε]ὺς κρ[ι] Η𐅄ΔỊỊ, | πυρῶν ΔΔỊỊỊ πένθ' ἡμιεκτεῖα· ἐξ Ἡφαιστίας στρατηγὸ[ς] Μνησίμαχος Ἁγνούσιος, Δημήτρ[ι]ος ἐκ Κοί, | Λυσιμαχίδης Ἁγνούσιος κρι ΗΗ𐅄ỊỊ ἑκτεὺς δύο χοίνικες, πυρῶν εἴκοσι τρεῖς δέκα ἡμιεκτεῖα δύ|[ο] χοίνικες· κεφάλαιον κριθῶν μέδιμνοι ΧΗΠỊỊỊ ἡμιεκτεῖα τέτταρα δύο χοίνικες, "From Skyros the general Mnesistratos of Kytheros [conveyed] forty-eight [*medimnoi*] of barley, eight [*medimnoi*] of wheat. From Myrina the general Sthenyllos of Eiresidai, Sopatros of Alopeke, and Aischylos of Erchia [conveyed] 162 [*medimnoi*] of barley, twenty-three [*medimnoi*] and five half-*hekteis* of wheat. From Hephaistia the general Mnesimachos of Hagnous, Demetrios from Koile, Lysimachides of Hagnous [conveyed] 252 [*medimnoi*], one *hekteus* and two *choinikes* of barley, twenty-three [*medimnoi*], ten half-*hekteis* and two *choinikes* of wheat. Total of the barley 1,108 *medimnoi*, four half-*hekteis*, and two *choinikes*."

This total figure of 1,108 *medimnoi*, four *hemiekteia*, and two *choinikes* in line 279 comprises all the barley from Attica, Drymos, Oropos, Salamis, Skyros, and the two cities on Lemnos (Myrina and Hephaistia). The Epistatai then record how much of this barley they used for sacred purposes, such as the *prokonia* and the *pelanos*, and note that they sold a specific quantity "at the price of three drachmai per *medimnos*, as set by the Assembly," πραθεισῶν ἐκ τριῶν δραχμῶν τὸν μέδιμνον ἕκαστον ὡς ὁ δῆμος ἔταξεν, lines 282–283.

The Epistatai next list in lines 283–284 the total number of *medimnoi* of wheat brought in as ἀπαρχή and repeat the process of reporting how much of it they used for the *pelanos* and for the *hieropoioi* from the *boule*, before noting that they sold a specific amount of this wheat at the price of six drachmai per *medimnos*, as set by the Assembly, except for ten *medimnoi* that yielded fifty drachmai, i.e., five drachmai per *medimnos*. They conclude with a grand total of money realized from the sale of the *sitos*, line 288.

After a list of expenditures in lines 288–297, the final entry on the stone concerned with grain reads as follows: ἐξ Ἴμβρο[υ τ]ῆς ἀπαρχῆς, οἱ ὕστερον (ἐ)λθόντες τῆς θυσίας, ἣν ἐκόμισε Χαιρέστρατος Ἀναγυράσιος, πυρῶ[ν] μέδιμνοι ΔΔΔΠΙ, δέκα ἡμ[ι]|εκτεῖα ἔχοντες τὴν ἐπιβολήν, τιμὴ τούτων ΗΗΔΔ𐅂, κριθῶν μέδιμνοι ΔΔΔΔΙΙΙ, ἡμιεκτε[ῖ]α τέτταρα ἔχοντες τὴν| ἐπιβολήν, τιμὴ τούτων, τοῦ μεδίμνου ἑκάστου 𐅂𐅂𐅂ΙΙΙΙΙ· σύμπαν τιμῆς κεφάλαιον ΗΗΗ𐅄ΔΔΠ𐅂𐅂 [.], lines 297–298. "Of the first-fruits from Imbros, the [*medimnoi*] for the sacrifice coming in late, which Chairestratos of Anagyrous conveyed, [consisted of] thirty-six *medimnoi*, ten half-*hekteis* of wheat, including the *epibole*, the price of these being 221 [drachmai]; forty-three *medimnoi*, four half-*hekteis* of barley, including the *epibole*, the price of these being three drachmai, five obols for each *medimnos*. Full total price 377[.] drachmai."

This remarkably well preserved record of the first-fruits from Attica and her possessions in 329/8 B.C. has traditionally been combined with the famous 5th-century B.C. First-Fruits Decree, *IG* I^3 78, ca. 425–415 B.C., in order to compute the production of wheat and barley in these regions in this year. The latter decree calls for 1/1200th of the wheat harvest and 1/600th of the barley harvest to be sent to Eleusis as ἀπαρχαί for the two goddesses, lines 5–7. Scholars

[66] Langdon (1987, p. 56, note 27) interprets this term as meaning not all of Oropos, but only that land belonging to the god, i.e., the coastal plain, whose fields produced the grain that he alone, apart from the Attic tribes, offered to the two goddesses. On the official in charge of this region, see Whitehead 1982.

have consequently multiplied the totals given in *IG* II^2 1672 by 1,200 and 600, respectively, to arrive at the total production of wheat and barley in Attica and her possessions in 329/8 B.C.[67] Wide-sweeping inferences have been drawn from such figures about the population of Attica, the sources of her food supply, the productivity of Attica compared to Lemnos, the role played by agriculture in shaping Athenian defense policy, the importance of barley, as opposed to wheat, in the Athenian diet, and many other things. A note of caution in method, however, is in order.

It was first sounded by Foucart (1884, pages 202–216), who questioned the veracity of the claim of continuity of Isokrates 4.31, in the sending of first-fruits to Eleusis by αἱ πλεῖσται τῶν πόλεων ca. 380 B.C., on the grounds that no foreign city is recorded on *IG* II^2 1672 as having sent grain to the goddesses in 329/8 B.C. He suggested that the 5th-century B.C. practice had lapsed and that procedures for handling the first-fruits as attested in this inscribed account of the Epistatai from Eleusis and the Treasurers of the Two Goddesses may have been a recent restoration of an ancient practice, perhaps introduced by the orator Lykourgos. Changes may have taken place in the meantime, and Foucart pointed to *IG* I^3 78, which orders collection of the *aparche* in Attica by the *demarchoi*, whereas in 329/8 B.C. donations from the same source are listed by tribes. The duties of the *hieropoioi* of Eleusis, prominent in the 5th-century B.C. decree, are in *IG* II^2 1672 all carried out by the Epistatai, and *hieropoioi* from the *boule* now appear. The two inscriptions differ also in the types of offerings made by the Epistatai and in the deities who received them, not to mention the considerably reduced quantity of the grain collected in the 4th century B.C. compared with what came into Eleusis in the 5th. It is possible that the proportions of the harvest had changed in the 4th century from 1/1200th for wheat and 1/600th for barley, although Foucart argued that something so traditional may well have remained unchanged.[68] He discounted a significant amount of cheating or misrepresentation in reporting the total harvest on the part of Attic farmers—in my view naively.[69]

We must be cautious, therefore, in extrapolating from procedures attested in the last third of the 5th century B.C. to those in effect in 329/8 B.C. On at least three occasions known to us in the interim, the Athenians legislated on the first-fruits at Eleusis. Changes in the procedure of collecting the *aparche* may have been introduced at these—or other—times.[70] With M. H. Hansen,[71]

[67] This process began with Tsountas 1883, cols. 259–262, in the editio princeps of *IG* II^2 1672; continued with Foucart 1884, pp. 211–216 and Gernet 1909, pp. 293–301; and found its most detailed treatment in Jardé 1925, pp. 36–60. Recent students of ancient agriculture who have relied heavily on these two inscriptions include Garnsey (1988 and 1992) and Sallares (1991).

[68] For some of these same reservations, see Jardé 1925, p. 42. Clinton (1994) argues "that 'many' Greek cities sent donations of first-fruits around the year 380" (p. 161). At the same time he cautions that "the position of Athens in 329/8 [*IG* II^2 1672] was far different from what it was when Isocrates wrote his *Panegyric*" (p. 170, note 1).

[69] More credible in my view is Jones 1957, p. 77: "I have no doubt that Athenian farmers all under-estimated their crops." So, also, Jardé 1925, pp. 42–43; Ober 1985, pp. 23–24. Gallo (1984, p. 61) urges that the "competenza dei demarchi . . . in un ambiente ristretto come il demo" would have made evasion of the first-fruits very difficult. Garnsey (1992, pp. 147–148) is willing to concede "some margin of error [in the extrapolated harvest figures of *IG* II^2 1672] due to evasion" but not enough to impugn the accuracy of those figures. According to Sallares (1991, p. 394), "We must also assume that farmers were reasonably honest in dedicating the appropriate proportion of their produce."

It is difficult to see how anyone can seriously maintain that tax evasion was insignificant in Classical Athens after the striking demonstration to the contrary by Cohen (1992, pp. 191–201). See also Burford 1993, pp. 93–95. The point will be important for us later; see below, pp. 113–114. For tax evasion and inaccurate reporting of harvests by Greek farmers in the 17th and 20th centuries, see Davies 1994 and Fotiadis 1995, pp. 72–76.

[70] (1) The law of Chairemonides mentioned in *IG* II^2 140, lines 9, 33–34; (2) the subsequent law in *IG* II^2 140 itself, 353/2 B.C.; (3) *SEG* XXX 61 (XXXVIII 57*), which in B fr. a, line 13 specifically mentions [τ]ῶν τὴν ἀπαρχὴν τõ σίτο ἀπαγόντ[ων] ("those rendering the first-fruits of grain"). For the view that regulations governing the computation and collection of the *aparche* may have been altered since the 5th century B.C., see Faraguna 1992, pp. 357–360.

[71] Hansen 1985b, pp. 24–25: "All attempts to calculate the size of the population of Attica from the annual consumption of grain, produced and imported, must in my opinion be rejected because (a) it is extremely difficult to calculate the amount of grain produced in Attica; (b) it is impossible to calculate the total import of grain; and (c) the calculations made so far have been based on a

therefore, I believe that all attempts to estimate the population of Athens on the evidence of *IG* II^2 1672 are seriously flawed in method.[72] Also, even if the inscription could provide reliable yield figures for wheat and barley in Attica and her possessions in 329/8 B.C., an isolated "statistic" for only one year is a shaky foundation on which to erect theories about the relative fertility of Attica and Lemnos, the amount of land under cultivation, the percentage of the Athenian population fed on local, as opposed to imported, grain, and similar topics.[73]

Another potential red herring in this entire exercise has been the claim that 329/8 B.C. was a particularly bad year. Doubts were raised by Jardé (1925, pages 43–47), but this has now become the standard view. Two representative supporters are Ober (1985, page 24): 329/8 B.C. was a "drought year which saw widespread crop failure" and Sallares (1991, pages 392–393). In my view the modern authorities cited, e.g., by Ober in his note 22 (page 24) do not substantiate this claim. It might be time, therefore, to scrutinize once more in detail all the ancient evidence for "widespread crop failure" in this year. Here I cannot do more than point out that the supporting evidence consists primarily of (1) a number of dated Athenian decrees in honor of people who sold grain at fair or cheap prices to the *demos* (collected in Pritchett 1991, pages 469–470; cf. Tracy 1995, pages 30–35) and (2) inflated prices of sacrificial victims, clothing, food, and other commodities in *IG* II^2 1672, especially when compared with those in the earlier *IG* II^2 1673 (Clinton 1971, pages 107–113).

Caution is required, however, in drawing sweeping inferences from the appearance of words denoting "grain shortage" in individual honorary decrees, as Pritchett (1991, page 470) has

---

consumption of wheat, whereas the sources indicate that the basic constituent of the Athenians' diet was barley." For similar caution see Foxhall and Forbes 1982, pp. 68–75: "Perhaps this study will best serve as a cautionary tale for researchers using grain consumption as one of the bases for constructing models of population size and/or structure, agricultural productivity, grain trade and other fundamental issues in classical social and economic history" (p. 75). See also Sallares 1991, p. 475, note 37. Cargill (1995, pp. 197–198) urges "extreme agnosticism" with regard to this inscription, observing that "the results are necessarily arbitrary, however laboriously that fact is disguised." He errs, however, in believing that *IG* II^2 1672 "has no entry for Imbros at all!" See lines 297–298.

A quick appreciation of the futility of modern guesses at the population of Rome ca. 5 B.C., where evidence is much more plentiful than for Athens, and at the numbers of *modii* necessary annually to feed it can be gained from, e.g., Gruen 1995, pp. 358–359 and Rickman 1980, pp. 8–13.

[72] Despite the best efforts of Gomme (1933a, pp. 28–35); Jones (1957, pp. 76–79); Gallo (1984, pp. 64–84); Ober (1985, pp. 22–27); and Sallares (1991, pp. 79–80).

[73] For striking variations in areas sown and yields of wheat and barley in only the two successive years of 1922 and 1923 in Greece, see Gomme 1933a, p. 31; for similar ups and downs in 19th- and early-20th-century France, see Jardé 1925, pp. 46–47; Gallant 1991, pp. 102–103. The perils of trying to estimate ancient yields of cereals by using modern comparisons are vividly exposed by Sallares (1991, pp. 372–389), whose discussion ought to induce extreme skepticism regarding this method. For the lack of evidence for accurately computing yields of grain in Italy, see Rickman 1980, pp. 103–104.

I have myself made no systematic examination of the productivity of the three islands in post-Classical times. The following random and impressionistic notices are, however, indicative of the futility of this method. Lemnos: 1860, with a population of 22,000 could produce 500,000 kg. of barley, of which about half was exported; Conze 1860, pp. 106–107, with a computation of the produce tax (φόρος ἐπιχαρπίας) paid in kind to the Turks. Skyros: April 18, 1702, "They have enough Wheat and Barley for their Subsistence: the French themselves come thither sometimes for these sorts of Grain" (de Tournefort 1741, II, p. 134); "The wheat of Skyro is equal to the best in the Aegaean. The productions are . . . 15,000 kila of corn, of which 2,000 are exported" (Leake 1835, III, p. 107), 500 families. In 1848, with a population of 2,630, Skyros produced ca. 40,000–45,000 kg. of wheat and barley annually (Graves 1849). In 1942, with a population of 3,179, "Skíros has a desolate and inhospitable appearance and it is, relatively, one of the least productive of the Aegean islands" (Naval Intelligence Division 1945, III, p. 406). See below, pp. 74–75. Imbros: in 1845, with a population of 8,000, the 10% produce tax in kind on σῖτος sent annually to Constantinople amounted to ca. 3,500–4,000 kg.; roughly the same amount was exported. Of κριθή the δέκατον amounted to 5,000–6,000 kg. and again roughly the same amount was exported from the surplus of the inhabitants, i.e., annual yield of σῖτος: 35,000–40,000 kg., of κριθή, 50,000–60,000 kg. (Moustoxydes 1845, p. 66). In 1978 Kalaitze (1989) records yields of roughly 32,850 kg. for wheat and 37,375 for barley, with 4,500 and 6,900 kg., respectively, exported.

warned: "The fact that the words σπανοσιτία or σιτοδεία occur [in a document] is in itself of little significance, and in no way justifies the term 'grain crisis' unless we apply it to all years. . . . Any assumption that there was no shortage if we do not have a spanositia text for a given year would be absurd." In the same vein, Gauthier (*BE* 1994, no. 441) has well remarked, "Je doute qu'on puisse établir une relation simple, dans une cité donnée, entre le nombre des décrets honorifiques gravés [for importers who offer grain at low prices] et le volume du grain importé." Cf. Gallo 1984, pages 61–69.

Consequently, I am not persuaded by, e.g., the repeated attempts of Garnsey (1985; 1988, pages 98–101, 158; and 1992, pages 148–149) to demonstrate on the basis of *IG* II² 1672 that in 329/8 B.C. Lemnos had an average or good harvest, while Athens had a poor one. Despite the optimism of Reger 1994 (page 110, note 71: "Garnsey makes an airtight case for good harvests on the islands and bad harvests in Athens"), his comparisons of the productivity of these two regions are based upon a "uniform figure for yield," which is unrealistic in view of the greater fertility of Lemnos. Sallares (1991, page 478, note 70) showed "why a 1 : 1 comparison between Athens and Lemnos is probably not justifiable."

Also, the high prices of animals and miscellaneous commodities attested for 329/8 B.C. may reflect—and indeed be caused by—"widespread crop failure." But it would be imprudent, in my view, automatically to infer that a shortage of food caused by drought was the sole determining factor in generating inflation. Many other factors could have been at work.

It may be true that the harvest of 329/8 B.C. was in some respects unusual in some places, but I do not find that the evidence as currently deployed gives us a very clear picture of the nature, the scale, and the potential geographical variation of the proposed grain shortage.

Other variables that seriously weaken previous attempts to draw inferences about production, population, and grain consumption from the figures of *IG* II² 1672 have been highlighted by Amouretti (1979); Gallo (1983); and Sallares (1991, pages 314–316, 366–368), who demonstrate, among other things, that Athenians consumed a great deal more barley than had hitherto been surmised. Foxhall and Forbes (1982) persuasively demonstrate also that traditional modern estimates of the nutritional (caloric) value of certain quantities of wheat and barley are woefully inadequate. Excellent on the fallability of *IG* II² 1672 as evidence for these and other aspects of Athenian agriculture is Garnsey (1992, pages 147–149): "no estimate of either total agricultural production, or the productivity of agriculture (and horticulture) in Attica can be drawn from the inscription." Ironically, while observing that "to inquire into the yield of the land in ancient Greece is to pursue a phantom" (page 147), he persists in presenting yet another tabulation of the estimated proportion of Attic land planted in wheat and barley with exact figures for yield, seed-yield, total output, and so forth. Even more elaborate and far-reaching reconstructions of yield, number of people fed, population density, estimated annual consumption, and the like, based largely on *IG* II² 1672, can be found in Gallant 1991, pages 177–179, and especially Sallares 1991, pages 79–80. Sallares is deaf to his own warning on page 394, "the information derived from it [*IG* II² 1672] is of limited value because we do not know whether it represents a good, bad or average year. A continuous run of data for twenty or thirty years would be needed to obtain statistically significant results for total production. However, if the information from the inscription is related to the proportion of Attica that is devoted to agriculture today . . . *it is an inevitable conclusion* [my italics] that the harvest recorded on the inscription was that of a below average year." In view of Sallares' frequent caustic attacks on what he regards as the puerile methods of classicists, ancient historians, and—worst of all—literary critics in interpreting written evidence, it is instructive to observe how a real scientist handles an ancient text.

It is time to recognize from the research of, for example, Foxhall and Forbes (1982); Gallo (1984); Garnsey (1988 and 1992); Sallares (1991); and Gallant (1991, pages 52–56, 60–78) that the more experts learn about the extreme variables in agricultural yield, nutritional value of different varieties of cereals, or the taxonomy of modern and ancient varieties of wheat and barley, the less reliable the single-year yield figures of *IG* II^2 1672 become, especially when they are juxtaposed to data from modern Greece. Isager and Skydsgaard (1992, page 70) have well observed, in another—but related—context, "the more sophisticated scientific methods become, the less inclined will the scholar be towards drawing a clear and unambiguous conclusion." Their analysis of "Agrarian Systems," pages 108–114, is a valuable exposé of theory and models running wild. It should be required reading for all students of ancient Greek agriculture. See Hodkinson 1988 for a timely warning about the lack of uniformity among Greek farmers in patterns of fallowing and rotation of crops.

Bearing all these caveats in mind, there might still be some minimal comparative value in the wheat and barley totals from Lemnos, Imbros, and Skyros on *IG* II^2 1672 for students of the grain-tax law. If the traditional *aparchai* proportions remained in effect in both 374/3 and 329/8 B.C., we might gain the limited information that in one year, forty-five years later than the date of Agyrrhios' law, Lemnos produced roughly 56,750 *medimnoi* of wheat and 248,525 of barley, while from Imbros 44,200 *medimnoi* of wheat and 26,000 of barley were harvested. Skyros produced 9,600 *medimnoi* of wheat and 28,800 of barley.[74] These totals are as close as we can get on present evidence to the amount of wheat and barley produced by the islands on which the Athenians imposed their tax of $8\frac{1}{3}$%. We return to this topic below, pages 41–42.

It is clear from the heading of Agyrrhios' law in lines 3–4, from the description of the tax to be farmed in lines 6–7, and from the formulation in lines 46–48, τὸν δὲ σ⟨ῖ⟩τον [ο]ἱ πριάμενοι τὴν δωδεκάτην κομισάντων, that the main topic of this inscription is an $8\frac{1}{3}$% tax on the grain produced in Lemnos, Imbros, and Skyros. As a second object of the verb πωλ[εῖ]|ν, however, in the clause ordering the selling of the *dodekate* (lines 6–8), we find another tax, καὶ τὴν πεντηκοστὴν σίτο. This 2% tax is not explicitly mentioned again until line 57 near the end of the law, where it is not qualified by σίτο. What is this tax and what is its relationship to the *dodekate*? Why was it not also written up in the heading in lines 3–4? Is it a tax that was assessed in cash or in kind? Is it a separate tax or is it to be added to the $8\frac{1}{3}$% tax? Is it a new tax or one already in existence?

Outside of the text of our law, I know of only one other explicit reference to such a tax at Athens. In [Demosthenes] 59.27, the speaker observes that Xenokleides, the poet (Kirchner, *PA* 11197), ἐωνημένος τὴν πεντηκοστὴν τοῦ σίτου ἐν εἰρήνῃ καὶ δέον αὐτὸν καταβάλλειν τὰς καταβολὰς εἰς τὸ βουλευτήριον κατὰ πρυτανείαν ("purchasing [the right to collect] the 2% tax on grain in peacetime and when it was incumbent on him to make his payments in the Bouleuterion each prytany"), in 368 B.C.[75] Clearly this 2% tax on *sitos* was collected in money. It was a tax "on" grain but not "of" grain, since Xenokleides had to make cash payments of the yield from this tax in the Bouleuterion each prytany.

Some scholars have identified this πεντηκοστὴ τοῦ σίτου as a tax of 2% on the value of all grain imported into Athens.[76] Efforts have been made to compute its annual yield by applying this percentage to the estimate of 800,000 *medimnoi* of imported grain each year derived from Demosthenes 20.31–33 in 355 B.C.[77] Others have equated it with the *pentekoste* farmed by Agyrrhios and Andokides in 402–400 B.C., which was an ad valorem tax imposed on all imports and exports in the Peiraieus (above, pages 19–20).[78]

[74] I have derived these totals from adding up the completely preserved numbers in lines 275–279, 297–298, as quoted above, pp. 32–34. Cf. also the helpful charts in Garnsey 1988, pp. 98, 101.

[75] Gernet's (1909, p. 368) attempt to link *IG* II^2 1707, line 6 (181/0 B.C.) to the πεντηκοστὴ τοῦ σίτου has now received some support from the new fragment added by Tracy (1984, pp. 370–374, no. 3); see *SEG* XXXIV 94; XXXIX 137*.

[76] Boeckh (1886, pp. 420–426), who distinguished it from the *pentekoste* on other commodities on the convincing grounds that it applied to imports only; see also Gernet 1909, pp. 367–368, who thought that it might go back to the 5th century B.C., and Lewis 1959, pp. 243–244, with the suggestion that the πεντηκοστὴ τοῦ σίτου "may have been a produce-tax on home-grown corn"; Gauthier 1981, pp. 6, 27; Hansen 1991, pp. 260–261.

[77] E.g., Boeckh 1886, p. 422.

[78] Garnsey 1988, p. 148. It is probable that the two isolated mentions of πεντηκοστή in a context of ναύκληροι, [ἔμποροι], ἐγγυηταί, ἀργύριον, and πράκτορες in *IG* I^3 133, lines 25 and 28 (later than the Kallias decrees)

In 20.32, however, Demosthenes states that written records of the quantities of imported grain were kept by the Athenian *sitophylakes*. Similar, but probably separate, records of the value of ships' cargoes exported from Athens were maintained by the *pentekostologoi* ([Demosthenes] 34.7). Such differentiation in record-keeping probably indicates that the πεντηκοστὴ τοῦ σίτου was farmed, collected, and recorded separately from the 2% tax on all other commodities imported and exported from Athens.

The case for identifying the πεντηκοστὴ τοῦ σίτου of [Demosthenes] 59.27 with the πεντηκοστὴ σίτο of line 8 in the present law is weak. First, by the terms of the opening sentence of this law, the πεντηκοστὴ σίτο is to be farmed for the specific purpose of making *sitos* available to the *demos* in the public domain. As we shall see, this means that whatever its prior status, the 2% tax will henceforth be assessed and collected in kind, not in cash. [Demosthenes] 59.27, however, shows clearly that the πεντηκοστὴ τοῦ σίτου in the Peiraieus was collected in cash. Second, payments of the latter tax were due in the Bouleuterion each prytany ([Demosthenes] *ad loc.*), whereas payments of the *dodekate* tax, to which the *pentekoste sitou* in line 8 is presumably assimilated, were governed by the festival calendar (line 48) and had to be paid at the Aiakeion (line 14), not the Bouleuterion. No attempt is made in the law to specify a schedule of payments (καταβολαί). The tax-farmers merely have to deliver the total number of *merides* before the month of Maimakterion. Third, the *dodekate* in line 6, which is likewise being sold to produce *sitos* for the Athenian *demos*, is defined as originating ἐν Λήμνωι καὶ Ἴμβρωι καὶ Σκύρω[ι]. It would not be too far-fetched to apply these locative prepositions also to the second object of the infinitive πωλ[εῖ]|ν, the *pentekoste sitou* in line 8. Hence, there seems to be no compelling evidence to suggest that Agyrrhios was here legislating about the πεντηκοστὴ τοῦ σίτου in the Peiraieus.

This reasoning would permit the conjecture that the *pentekoste sitou* in line 8 is a hitherto unknown tax, one that, like the *dodekate*, was levied not in the Peiraieus but in the three islands. At the Peiraieus the *sitos* thus taxed consisted of cargoes of grain imported into Athens. Since, as our law now shows, the three islands all exported wheat and barley to the mother city, it is possible that their 2% tax *sitou*, if that is what it is, was levied on the cargoes of grain that left the harbors of the three islands. The most plausible explanation of Agyrrhios' instructions in lines 6–8 to "sell" the *pentekoste sitou* is that it too was auctioned in Athens. Taxes collected in the three islands would then be administered from Athens, as was the πεντηκοστὴ ἐν τῆι Νέαι, above, pages 28–31.

In seeking to refine our definition of τὴν πεντηκοστὴν σίτο in line 8, we may perhaps gain additional insight from the context of the law itself, rather than from apparent verbal parallels such as ἡ πεντηκοστὴ τοῦ σίτου of [Demosthenes] 59.27. The sole passage in which a πεντηκοστή recurs is lines 55–59, where the *apodektai* are instructed to allocate (μερίσαι) a portion of its cash revenue to the general fund (διοίκησις). I shall return to the discussion of what can be inferred about the status of this tax and about the purpose of the obscure financial maneuvers in lines 55–61 of the law; below, pages 78–84. To anticipate future conclusions, I will there argue that the 2% tax in lines 8 and 57 are one and the same.

For present purposes it is enough to note that in line 57 τῆς πεντηκοστῆς is not qualified by the word σίτο, as it appears to be in line 8. This may simply be due to careless drafting on the part of the lawgiver or perhaps he surmised that all readers would make the obvious connection with τὴν πεντηκοστὴν σίτο in line 8, thereby rendering repetition of the qualifying noun superfluous.

It is worth exploring, however, a further implication of the "omission" of the word σίτο as a qualifier of the *pentekoste* in line 57. Its absence would economically leave us with a simple 2% tax that most plausibly could be identified as a levy on all harbor activity in the islands not

designate this 2% tax. Like most Athenian taxes, it merits careful restudy in the light of subsequent discoveries; see above, pp. 27–28. There is a good general discussion of this tax in Velissaropoulos 1980, pp. 208–211; cf. Gauthier 1981, pp. 6, 27; Gofas 1993, pp. 225–227.

restricted to grain—similar to the πεντηκοστὴ ἐν τῆι Νέαι and indeed to the familiar 2% tax in the Peiraieus farmed by Agyrrhios and Andokides, above, pages 19–20. As in these cases, the πεντηκοστή in line 57, at least before 374/3 B.C., would have been collected in cash in Lemnos, Imbros, and Skyros.

If the tax in line 57 is in fact the same tax as τὴν πεντηκοστὴν σίτο of line 8, we seem to be presented with the awkward challenge of dissociating the qualifier from the tax in line 8 and rejecting the translation "the 2% tax on grain." Here it may be helpful to note that in our only known Athenian verbal parallel, ἡ πεντηκοστὴ τοῦ σίτου, [Demosthenes] 59.27, the definite article is used with grain. Similarly, in our law we find περὶ τῆς δωδεκάτης τοῦ σίτου, line 3. For what it is worth, the definite article is also found with σῖτος in eleven other passages in our law. In only two places does σῖτος appear without it, lines 5 and 8. In line 5 its omission is explicable: ὅπως ἂν τῶι δήμωι σῖ[το]|ς ἦι ἐν τῶι κοινῶι. The grain here is not specific grain, as for instance that in περὶ τῆς δωδεκάτης τοῦ σίτου| τῶν νήσων, lines 3–4, or in [κο]|μιεῖ τὸν σῖτον, lines 10–11, etc. It is simply grain as a commodity: "In order that there may be grain for the people." Could this also be how we should interpret σίτο in line 8?

Two indicators point in that direction. First, in lines 8–10, immediately after instructions to sell the *dodekate* and the *pentekoste* in order that the people will have *sitos* in the public domain, the law moves abruptly to a definition of each *meris* in terms of wheat and barley. Clearly we are swiftly in the context of a tax that will be collected in kind, not in cash. But despite the clearly stated purpose of the law, how would we have been prepared for this step if the instructions merely stated, "sell the $8\frac{1}{3}$% tax in Lemnos, Imbros, and Skyros and the 2% tax on grain"? Confusion might have arisen in the minds of some readers between the πεντηκοστὴ σίτο of line 8, which was clearly supposed to produce grain, and the πεντηκοστὴ τοῦ σίτου in the Peiraieus, which was collected in cash.

Second, to anticipate later discussion, the 2% tax in line 57 had clearly been collected up to the time of our law in terms of cash. If we identify it with the πεντηκοστὴ σίτο in line 8, as I think we must, then how could instructions in lines 6–8 to sell a 2% tax on grain, previously collected in cash, be expected to produce *medimnoi* of wheat and barley as specified in lines 8–10? I suggest that the status of this 2% tax is being changed by our law from a tax collected in cash to one collected in kind.

It is very doubtful, however, that Athenian readers would have been required to go through this intricate progression of inferences to reach this conclusion. It is also unlikely that Agyrrhios, in making what seems to have been a momentous shift in the method of collecting these two taxes, would have omitted explicit reference to this point and simply have moved from instructions to sell the *dodekate* and the *pentekoste sitou*, directly into specifications of *merides* in terms of *medimnoi*.

If σίτο stands alone in the strong position at the end of its clause and does not depend upon τὴν πεντηκοστήν, then it could serve the double purpose of making explicit the most important innovation of Agyrrhios' law and of providing a transition to the definition of *merides* in the form of wheat and barley which follows it. It then becomes a pivotal word, explaining how what precedes is going to put *sitos* into the hands of the people, and why the specifications of the portions immediately follow it. As a kind of genitive of material, it makes explicit the terms or the commodity in which both taxes will henceforth be collected. As a kind of partitive genitive, it helps define the whole of which the μερίδες in lines 8–10 are parts. This may be to load too much weight onto one small word, but it is worth considering the possibility of a translation such as, "In order that there may be grain for the people in the public domain, sell the $8\frac{1}{3}$% tax (which originates) in Lemnos, Imbros, and Skyros and the 2% tax in terms of grain. Each portion (of it) will consist of 500 measures (*medimnoi*)."

LINES 8–10

ἡ δὲ μερὶς ἐχ̣[ά]-
στη ἔσται πεντακόσιοι μέδιμνοι, πυ[ρῶ]-
ν μὲν ἑκατόν, κριθῶν δὲ τετρακόσιοι

Immediately after the instructions to "sell the *dodekate* and the *pentekoste* in terms of grain," the law proceeds to define each μερίς as consisting of 500 *medimnoi*, 100 of wheat and 400 of barley. As ἐχ̣[ά]στη here makes clear—and κατὰ τὴν [μ]ερ[ίδ]α in lines 28 and 30—Agyrrhios had many μερίδες in mind. In lines 31–36 a different, much larger, μερίς appears—one consisting of 3,000 *medimnoi* shared among a *symmoria* of six men. Both types of μερίς, however, are defined in terms of grain. Μερίς normally designates a "portion" or a "share." The word appears most frequently in surviving Attic inscriptions as a portion of sacrificial meat distributed to participants in festivals[79] or as separate sections of fortification walls for which different teams of construction workers contract with the state to make repairs.[80] Although the term is not common in contexts of taxation in the Classical period, I take it to mean, in Agyrrhios' law, a portion of grain, consisting of 500 (or 3,000) *medimnoi*, which the farmers of the *dodekate* and the *pentekoste* contract with the state to deliver to Athens. The law says nothing about how many μερίδες an individual tax-farmer was permitted to "buy." It does not say, however, and perhaps may not imply, "only one μερίς per customer." Provided he could pay the twenty drachmai fee and produce the two solvent guarantors required per μερίς (lines 28–31), a tax-farmer might have been able to contract with the *polis* to deliver more—perhaps many more—than one "portion" of the people's grain to the Aiakeion. On this point see also the discussion on pages 115–116 below.

It is significant, I think, that once the tax-farmers deliver the wheat and barley to Athens and weigh it out under supervision in the Aiakeion, the law shows no further concern with μερίδες. They belong to the process of farming out and collecting the tax and transporting the grain from the islands to Athens. Once the "ten men" (lines 36–44) take charge of the public grain in the public granary, μερίδες disappear. When they turn over their μερίδες of grain on time and in the prescribed condition, the πριάμενοι of the taxes have, presumably, discharged their contractual obligations with the state.

The law nowhere specifies how many μερίδες of grain the Athenians intend to "sell" to the tax-collectors in the first year of its operation, or in subsequent years. At the time of each annual

[79] E.g., law on the Lesser Panathenaia, 336–334 B.C., *IG* II² 334, lines 10–13: νείμαντ[ας τοῖς πρυτάν]εσιν πέντε μερίδας καὶ τοῖς ἐννέα ἄρ[χουσιν τρεῖς] καὶ ταμίαις τῆς θεοῦ μίαν καὶ τοῖς ἱερ[οποιοῖς μίαν] ("distributing to the prytaneis five portions and to the nine archons three and one to the treasurers of the goddess and one to the hieropoioi"), lines 25–27: ἀ[πονέμειν δὲ] τὰς μερίδας εἰς τὸν δῆμον ἕκαστον κατὰ [τοὺς πέμπον]τας ὁπόσους ἂν παρέχηι ὁ δῆμος ἕκαστος ("allocate the portions to each deme in accordance with the numbers of participants in the *pompe* that each deme provides"); Deme decree of Eleusis, ca. 350 B.C., *IG* II² 1187, lines 20–23: νέμειν δὲ αὐτῶι καὶ μερίδα ἐκ τῶν ἱερῶν καθάπερ Ἐλευσινίοις τὸν δήμαρχον ("let the demarchos allot to him also a portion from the sacrifices just as to the Eleusinians"); Deme decree of the Peiraieus, ca. 300–250 B.C., *IG* II² 1214, lines 11–14: ὅταν θύωσι Πειραιεῖς ἐν τοῖς κοινοῖς ἱεροῖς νέμειν καὶ Καλλιδάμαντι μερίδα καθάπερ καὶ τοῖς ἄλλοις Πειραιεῦσιν ("whenever the men of the Peiraieus sacrifice in their common sanctuaries, allot also to Kallidamas a portion just as in the case of the other men of the Peiraieus also"); cf. also *IG* II² 1231, lines 9–13; 1254, lines 10–12; 1330, lines 32–33; *SEG* XXI 469, line 55; in general see Detienne and Vernant 1989, pp. 13, 213, no. 11 (bibliography).

[80] Cf. the contract for construction of the walls of Athens and the Peiraieus, 307/6 B.C., *IG* II² 463, lines 117–122, κατὰ τάδε ἔνειμαν οἱ [ἀ]ρχιτέκ[τονες τ]ὰ μέρη τοῦ τείχους· πρώτ[η] μ̣ερὶς τ̣ο̣ῦ̣ ν̣[οτίου τεί]χους *vacat* κατὰ τάδε μεμίσθωται τὰ ἔρ[γα τὰ πε]ρὶ τὰ τείχη εἰς τὴν τετραετίαν· [τοῦ β]ορείου τείχους πρώτη μερὶς [ἀπὸ τ]οῦ διατειχίσματος μέχρι τῶν [πρώτω]ν πυλῶν ("The architects have distributed the portions of the wall in the following manner: first *meris* of the south wall *vacat* The works on the walls for the four-year period have been leased out on the following terms: Of the north wall first *meris* from the crosswall up to the first gates"). See Maier 1959, pp. 66–67. For μερίδες of arable land, see *IG* XIV 645, I, lines 14–30.

auction of the $8\frac{1}{3}$% and 2% taxes, however, the *bouleutai*, the *poletai*, and potential bidders must have been able to estimate roughly how much wheat and barley these taxes were likely to produce. Otherwise, they would not have known how many μερίδες were likely to come into play in the auction and how much storage space was required in Athens for the people's grain. Since the total number of "portions" could presumably fluctuate from year to year, depending on the amount of grain available and on the state's ability to attract viable bidders, we should probably conclude that determination of the annual total was left to the discretion of the *bouleutai*, possibly in consultation with the *poletai* and others.

Our own curiosity as to the quantities of wheat and barley the lawgiver had in mind can receive perhaps only partial satisfaction if we turn again to the yield figures extrapolated from the Eleusinian *aparchai* inscription of 329/8 B.C. considered above, pages 32–35. These totals from more than forty years later than the publication of Agyrrhios' law must be used with great caution, as we have seen. They permit, however, the following rough estimates (all totals in *medimnoi*):

	LEMNOS		IMBROS		SKYROS	
	wheat	barley	wheat	barley	wheat	barley
Extrapolated total yield	56,750	248,525	44,200	26,000	9,600	28,800
$8\frac{1}{3}$%	4,729	20,710	3,683	2,166	800	2,400
No. of μερίδες	47	52	37	5	8	6

It is obvious that the number of *medimnoi* equivalent to $8\frac{1}{3}$% of the total yield figures from 329/8 B.C. cannot easily be divided up into neat μερίδες of 100 *medimnoi* of wheat and 400 of barley such as Agyrrhios prescribed in his law of 374/3 B.C. We can only work out some crude estimates, and I have not included the 2% tax in the above calculations. In Lemnos the situation is fairly straightforward, for the two figures of forty-seven μερίδες for wheat and fifty-two for barley are close enough that we might settle on a figure of about fifty μερίδες from this island. In Imbros, however, which in 329/8 B.C. apparently produced more wheat than barley, we have totals that translate into thirty-seven μερίδες of wheat but only five of barley. The discrepancy here is so wide that one hesitates to propose even a rough total of μερίδες. The figures from Skyros yield eight μερίδες for wheat and six for barley, so that we could estimate a total of seven μερίδες from this island. Clearly only rough approximations are possible. At the lowest end of the scale, we would have 50 (Lemnos) + 7 (Skyros) = 57 total μερίδες as an unrealistic minimum. Even on the most conservative estimate of the situation in Imbros, we could propose a minimum of five μερίδες, which would increase the total to sixty-two. But clearly Imbros was more productive than Skyros and the total number of μερίδες it produced must certainly have been higher.

If, then, we could trust this comparison with the extrapolated yield figures of grain from the three islands in 329/8 B.C., the Athenians in 374/3 B.C. might have expected to auction off roughly a minimum of ca. sixty-two μερίδες and to have ready for sale in the Aiakeion at least about 6,200 *medimnoi* of wheat and 24,800 of barley. As we shall see later (below, page 97), calculated on the basis of the carrying capacity of a normal merchant ship, the total would be equivalent to about 6,200 + 24,800 = 31,000 divided by 3,000 *medimnoi*, which works out to a little more than ten shiploads of public grain to be transported to the Aiakeion and later sold to the *demos*. Again, we must emphasize that this is only a very crude estimate, using the only evidence we have available. It is very probably quite on the low side, especially since estimates of the yield of the *pentekoste* have not been included .

The proportion of 100 *medimnoi* of wheat to 400 of barley, specified in lines 9–10 for each μερίς, provides potentially valuable new information for students of ancient agriculture.[81] We note first

[81] It is important to remember that the proportion here is between wheat and barley (κριθαί) and not between wheat flour and barley meal (ἄλφιτα). See below, pp. 42–43. The latter comparison forms the basis of almost all modern attempts to determine the relative weights, rates of consumption, and nutritional and monetary values

that it does not correspond to the wheat : barley ratio reflected in the total yield figures from the three islands extrapolated from *IG* II^2 1672 for 329/8 B.C. In the latter, the overall ratio of wheat to barley is 1 : 3.004.[82] Even more at odds with the proportions in our law is the wheat : barley total yield ratio for Attica extrapolated from this same source, which Sallares (1991, page 314) has calculated as 1 : 9.3. Such discrepancies should probably warn us yet again that these later first-fruits figures from Lemnos, Imbros, and Skyros may be misleading.

Prior to the discovery of our law, scholars had repeatedly turned to literary and epigraphical sources and sometimes—with disastrous results—to parallels with grain production in modern Greece[83] in attempts to estimate the proportion of wheat to barley in ancient yield and consumption figures. In my view, many of these efforts have led to unsatisfactory results because their proponents set their sights too high. They seek global figures, "normal" or "average" yield and consumption totals, something that they can apply all over the ancient Greek world. Almost every other aspect of the study of ancient Greek history, however, teaches us the importance of regional variation. It is faulty method, in my view, to extrapolate yield or consumption figures of wheat and barley from one ancient *polis* to another, just as it is to do so with laws, cults, weights and measures, alphabets, coinage, or water supply. One has always to look very carefully at the local data and often to be content with the specific, restricted inferences they permit. This is one of the valuable lessons to be learned from Reger's excellent 1993 study of the public purchase of grain on Delos: "Large consuming centers like Athens, which depended on imports from the Black Sea, Egypt, and the Western Mediterranean, were structurally different from small centers like Delos, plugged into a regional market." Even though Delos has produced exceptional epigraphical data on grain prices, "the experiences of the one cannot be transferred to the other" (page 331). See also Reger 1994, pages 83–126, for an examination of the local peculiarities of the Delian grain trade.

In the present case, *IG* II^2 1672, we have a striking discrepancy between wheat : barley ratios of total yield figures of 1 : 3 for Lemnos, Imbros, and Skyros in 329/8 B.C. and 1 : 9.3 for Attica. Faced with such disparate figures, I do not believe that wide-ranging inferences, such as assuming a bad year in Attica and a good one on Lemnos or the like are justified. If, on the other hand, we look hard at the figures we have in the light of the same kind of regional differentiation that permeates the world of the Greek city-states, we could reach the perhaps not so startling conclusion that in 329/8 B.C. Attica, for some reason, produced a great deal more barley than wheat in comparison with her three island possessions. For all we know, this might be the norm. If we go farther and break down the overall ratio for the latter into its three regional components, a more detailed and complex picture emerges of Lemnos at a ratio of 1 : 4.37, Skyros at 1 : 3, and Imbros with a reverse ratio of barley to wheat of 1 : 1.7. The corresponding wheat : barley ratio for Attica, as we have seen, is 1 : 9.3.

Extrapolations of ratios from these production figures for only one year, at first glance, do not appear to shed much light on the proportion of wheat and barley consumed by the Athenians who helped pass Agyrrhios' law, except perhaps to confirm the demonstration by much recent research that Athenians ate a great deal more barley than had previously been supposed.[84] In this connection, Foxhall and Forbes (1982, page 67) have aptly observed, "Although a considerable

---

of these two commodities in antiquity. For a helpful discussion see Foxhall and Forbes 1982, pp. 51–65, with the correction of Gallo (1983, p. 469, note 80); cf. Gallo 1984, pp. 31–42; Sallares 1991, pp. 367–368; Gallant 1991, pp. 60–75.

[82] See above, p. 41: 110,550 *medimnoi* of wheat; 332,125 of barley.

[83] The perils of pressing these modern "parallels" too far are well illustrated by the examples studied, e.g., by Foxhall and Forbes (1982, pp. 65–72); Gallo (1983, pp. 463–464, note 27); Garnsey (1988, pp. 10–14); Sallares (1991, pp. 313–316); and Garnsey (1992, p. 149). Especially cogent are the warnings issued by Fotiadis (1995) about using "statistics" on population and agricultural productivity in modern Greece.

[84] E.g., Pritchett 1956a, pp. 185–186; Gallo 1983; Gallo 1984, pp. 23–35; Hansen 1985b, pp. 24–25, quoted above, pp. 34–35, note 71; Garnsey 1988, pp. 51, 99; Sallares 1991, pp. 313–316, 367–368; Pritchett 1991, p. 472, note 703; Reger 1993, p. 304, note 13.

amount of barley was eaten in ancient Greece, we have no evidence from antiquity for what proportion of barley in relation to wheat was eaten. Even the classical grain distribution figures are in *either* barley meal *or* wheat. The amount of barley eaten probably varied immensely, both regionally and with class and income level." Our new law might help a little here. We must be cautious, however, not to exaggerate the importance of the valuable new evidence for the proportion of wheat to barley of 1 : 4 that lines 9–10 of Agyrrhios' law now provide. This proportion is what one Athenian lawgiver in 374/3 B.C. hoped to produce in the public granary for sale to the *demos* by his tax levied on three specific—and perhaps very different—islands. We must not leap on this ratio, therefore, and expand it into a global estimate for wheat and barley consumption rates in the rest of Greece. It may not even be a valid long-term ratio for consumption in Attica.

And yet Agyrrhios clearly expected the "selling" of the $8\frac{1}{3}$% and 2% taxes to result in four times as many *medimnoi* of barley coming into the Aiakeion as wheat. How did he settle on this ratio? Did he simply take the proportion of wheat to barley grown on Lemnos, the largest and most productive of the islands, and apply it to the other two? The proportion of wheat to barley from Lemnos in 329/8 B.C. is suggestively 1 : 4.37, and there is a slight possibility that Agyrrhios served in an official capacity on that island at some earlier date.[85] In view of his own personal experience as a tax-farmer and his prior involvement in the importation of grain to Athens,[86] however, he may have arrived at this ratio after more informed and perhaps sophisticated calculations.

We would also love to know why the lawgiver set the size of each *meris* of grain at 500 *medimnoi*. As we shall see later, this figure is roughly one-sixth the normal carrying capacity of a merchant ship in the Classical period, and this fact seems clearly to have been in Agyrrhios' mind when he drafted the regulations in lines 31–36 on the expanded *meris* of 3,000 *medimnoi*, distributed among a *symmoria* of six πριάμενοι. For most modern readers, however, the words πεντακόσιοι μέδιμνοι in line 9 will almost inevitably resonate with Solon's τέλος of οἱ πεντακοσιομέδιμνοι. This may be to read too much back into our exiguous sources. We have been told that in 4th-century B.C. Athenian democracy the *pentakosiomedimnoi* "existed as a mere formality with no significance" (Hansen 1991, page 109).[87] But this same scholar's own careful research has revealed that the old Solonian τέλη still retained some practical applications in the first decades of the 4th century B.C.[88] Without going into the vexed, intertwined problems of the origin of the "500 measures" and the methods for computing them, I find attractive W. R. Connor's (1987, pages 47–49) suggestion that the four Solonian τέλη may have been closely related to the collecting, public display, and dedication of first-fruits of the wheat and barley harvest. If this system had been "sanctioned by long custom and religious usage," as he proposes, there might still have remained in 374/3 B.C. something basic and standard about a quantity of grain consisting of 500 *medimnoi*.

It is noteworthy that the lawgiver defines the wheat and barley that the *priamenoi* ship to Athens in terms of a measure of capacity—the *medimnos*—and not one of weight. Later, in the Aiakeion, the people's representatives will supervise the weighing of the people's grain and check its quality (lines 21–26), and we learn in fact that a *medimnos* of barley will have to weigh one talent. But the fact that the size of a *meris* of grain is given twice in our law in terms of *medimnoi* lends some support to Wallinga (1964, pages 6–9), who argued against the view that in Classical Athens there was a uniform standard of measuring a ship's capacity by weight—in terms of talents. He opts persuasively for a variety of methods of measuring a ship's capacity depending on the cargo; the cargo of a grain ship was normally reckoned in *medimnoi*.

[85] See above, p. 22.

[86] See above, pp. 19–20, 22.

[87] Cf. also Thomsen 1964, pp. 110–111. Rhodes (1981, pp. 137–142) collects most of the evidence for the *pentakosiomedimnoi*.

[88] Rhodes 1981, pp. 43–46, 106–109. The latest attestation of πεντακοσιομέδιμνοι in an Attic inscription known to me is (significantly?) the decree concerning Lemnos *IG* II^2 30, line 12, 387/6 B.C., where the context is uncertain; see Stroud 1971, pp. 162–173, no. 23; *Agora* XIX, L3; Figueira 1991, p. 61, note 36. For a good discussion see Hornblower 1991, pp. 399–400.

Agyrrhios' law breaks down *sitos* into its two most common components, wheat, πυροί, and barley, κριθαί, both expressed in the plural throughout. See above, page 16. For a full and illuminating discussion of the varieties of wheat in ancient Greece with fascinating observations on its evolution, the difficulties of precise identification, the dangers of modern "parallels," and seed-mixing in antiquity, see Moritz 1958, pages xvii–xxvii, 146–150, 168–176; see also Sallares 1991, pages 313–361, with copious bibliography.

The law instructs the tax-farmers to bring barley, κριθαί, not barley meal, ἄλφιτα, from the islands to the Peiraieus and up to the Aiakeion in the *asty*. It was, then, still a whole grain, and not ground into flour or meal. As far as we can tell from the text of the law, it remained in this state throughout the entire process of the heaping up (line 14), the weighing (lines 23–26), the quality control for dampness and darnel (lines 24–25), and the storage for a minimum of three months in the public granary (lines 42–48). There is no indication that the whole barley from the islands was ever ground while it was under the care of the ten men. In fact it is clear that these officials sold it as whole grain, and not as barley meal, at the end of the procedure authorized in the law, for when the demos finally set the selling price, the grain was still designated as τῶν κριθῶν in line 45. For similar public sales of σῖτος, not flour, see Migeotte, forthcoming, "Ventes."

After using the infinitive πωλ[εῖ]|ν as imperative to express his instructions to sell the *dodekate* and the *pentekoste* in the first main clause of the law (lines 6–8), Agyrrhios interestingly shifts to the future indicative, ἔσται, in line 9 in the second clause where he defines the composition of a μερίς. Syntactically, this clause thereby achieves a certain degree of independence, because its finite verb is not strictly dependent on εἶπεν in line 5. That this was not a slip or an isolated stylistic quirk is clearly demonstrated by the remarkable string of twelve consecutive future indicatives that follow ἔσται in line 9 and monopolize the syntax in all the main clauses of the law down to line 36. This continuity of tense and mood survives six different changes in subject as Agyrrhios moves back and forth from κομιεῖ . . . ὁ π̣[ρ]ιάμενος (lines 10–12 with asyndeton) . . . καὶ ἀνακομι[ε]ῖ (lines 12–13) . . . καὶ κατανήσει (line 14) to παρέ[ξ]ει . . . ἡ πόλις (lines 15–16) back to ἀποστήσει . . . [ὁ] πριάμενος (lines 16–18) to οὐ πράξει [ἡ π]όλις (lines 20–21) back to ἀ̣[πο]στήσει ὁ πριάμενος (lines 21–22 with asyndeton) and ἀπο̣σ[τ]ήσει (lines 24–25) . . . οὐ θήσει ὁ πριάμενος (line 27 with asyndeton) . . . καταστήσ[ε]ι ὁ πριάμενος (lines 29–30 with asyndeton) to συμ[μορ]ία ἔσται ἡ μερίς (lines 31–32 with asyndeton) and finally to ἡ πόλις πράξει (line 33 with asyndeton).

In line 36, the next clause begins with an imperative introducing a totally new subject, which is not ὁ πριάμενος, ἡ πόλις, or ἡ μέρις, but αἱρείσθω δὲ ὁ δῆμος. From here until the end of the law, the future indicatives disappear, except in the dependent relative clause οἵτινες ἐπιμελήσονται containing instructions for the "ten men" in line 39. After imperatives take over in line 36, they in turn monopolize the syntax of the next nine main clauses of the law—except for the isolated infinitive μὴ ἐ[ξε]ῖναι in lines 42–43—down to line 55.[89] Here a new construction takes over and to express his instructions to the *apodektai* in the final six lines of his law, Agyrrhios employs only infinitives, μερίσαι (line 56), εἶναι (line 59), and μὴ ⟨ἀ⟩φαιρεῖν (line 60).

Such variety of construction is very unusual in Attic public inscriptions and, to my knowledge, the consecutive string of thirteen finite verbs in the future indicative is completely without a parallel in the main body of an Athenian law.[90] Striking also is the heavy concentration of asyndeton. Six of the thirteen verbs lack any form of connection. In Nikophon's closely contemporary law on silver coinage (*SEG* XXVI 72), by comparison, imperatives dominate the syntax throughout, together with a few infinitives, and there is not one case of asyndeton. All other surviving Athenian laws on stone couch their instructions in either imperatives or infinitives, or a combination of

[89] On the form of the third person imperatives in -ντων and -σθων here and in lines 47, 49, 52, 53, see Threatte 1996, pp. 462–466; Harris 1992, p. 76.

[90] In the law of 353/2 B.C. regarding the first-fruits at Eleusis, *IG* II^2 140, there are two consecutive future indicatives in lines 25 and 27.

both; see the list of these laws above on pages 15–16. Nor have I found valid parallels for this concentration of future indicatives in the main text of decrees from Attica.

It is significant, I think, that Agyrrhios employs the future indicative only in that section of his law which contains regulations concerning the πριάμενος/οι and the πόλις. Apart from the definition of a μερίς in lines 9 and 32, either one of these two parties to the agreement regarding the farming out of the *dodekate* and the *pentekoste* is the subject of all the future indicatives. In lines 31–36 the συμμορία stands in for the πριάμενος/οι. When he shifts his construction to the imperative in line 36, Agyrrhios turns his attention to the δῆμος and to the δέκα ἄνδρες who now become the subjects of almost all the main clauses in the law down to line 55. The πριάμενοι make only one brief reappearance as subject of the imperative κομισάντων in line 47, but it is clear that in this section of the law the focus is now upon the δῆμος and its representatives, the ten men.

Another striking feature of lines 36–55, after the string of future indicatives, is the lawgiver's abandonment of asyndeton or—to put it positively—in contrast to the preceding section, lines 5–36, in which only three of the eleven main clauses contain a connective particle, each of the seven clauses in lines 36–55 is linked to its predecessor by δέ. This is also true in the third and final section of the law, lines 55–61, where the imperative infinitives are introduced by δέ and by τὸ μὲν νῦν . . . καὶ τὸ λοιπόν.

A more detailed stylistic analysis of this inscription than I have time for here would probably reveal other interesting features, for instance, that with only two exceptions (lines 8–9 and 33) all of the subjects of the future indicatives in lines 5–36 are written after the verb, whereas the subjects of six of the eight imperatives in lines 36–55 come before the verb.[91] Also worth comment is the considerable variety (μεταβολή) in the sentence structure and word order throughout. Note too the frequency of parallel structure, the fondness for placing the most important word first in the sentence, especially when it is the object (lines 6, 20, 21, 27, 46, 55), and the almost chiastic arrangement of lines 16–21. Clearly, the discovery in the Agora Excavations of Nikophon's law on silver coinage (56 lines), Agyrrhios' law (61 lines), and the substantial text of the law on Agora I 7495 (ca. 32 lines), which J. McK. Camp II and M. B. Richardson will publish, now makes available a significantly increased body of material on which to base a detailed study of the prose style of inscribed Athenian laws in the 4th century B.C. See the list of laws above on pages 15–16. Also awaiting closer analysis is a comparison of the style and syntax of inscribed Athenian laws with that of contemporary decrees.[92]

In the context of inscribed Athenian laws, Agyrrhios' thirteen consecutive finite verbs in the future indicative in lines 9–36 find their closest parallel in the law of 337/6 B.C. on the rebuilding of the walls in the Peiraieus, *IG* II^2 244. Instructions in the main body of this enactment, lines 1–46, are expressed uniformly by infinitives. Appended to the law, however, and originally inscribed in four narrow columns below the main text, are [Σ]υγγραφαί, of which only parts of the first two columns at the left have survived, lines 47–113. These contain instructions to οἱ μισθωσάμενοι

[91] In Nikophon's law on silver coinage, *SEG* XXVI 72, twelve imperatives are preceded by their subjects and nine others are written before their subjects. Two other imperatives are impersonal, ὑπαρχέτω followed by the dative, lines 30, 35.

[92] This is a rich and almost untouched field for future research. Although these texts have been studied for their grammar and formulas, especially by scholars like Larfeld, Meisterhans, Threatte, and Henry, little has been done to analyze their style. See the valuable and characteristically provocative observations of Dover (1981), where the emphasis falls mostly on diction. Except for the excellent pp. 13–15, Dover (1997) has little to say about this aspect of Attic inscriptions in his fundamental study of the evolution of Greek prose style. T. L. Shear, Jr., reminds me of another recent substantial contribution of the Agora Excavations to the growing body of material available for further stylistic analysis in the long (109 lines) and well-preserved honorary decree for Kallias of Sphettos of 270/69 B.C., with its remarkably frequent use of the genitive absolute construction, *SEG* XXVIII 60.

Some idea of the value of such a study for our understanding of the development of Greek prose style may be gained from Bloch 1975. See also the important stylistic discussion of Drakon's law on homicide by Gagarin (1981, pp. 153–161).

τὰς τομὰς τῶν λίθων ἐπὶ τὰ τείχη, which are first expressed in four consecutive finite verbs in the future indicative, lines 47–66. All are linked to one another by καί. In lines 66–98, the construction shifts to infinitives, five of them, only to return to three more consecutive future indicatives, this time with asyndeton, lines 98–113, before the stone breaks off. Other Athenian συγγραφαί of the 4th century B.C. show more future indicatives; for instance, in the specifications for the Arsenal of Philon, 347/6 B.C., *IG* II^2 1668, after an initial infinitive, there are forty-seven consecutive main verbs in the future indicative (with connective particles); see also *IG* II^2 1670–1671, ca. 330 B.C.; 1675, 337/6 B.C.; 1678, before 315 B.C.; 1685, ca. 300 B.C. In the συγγραφή preserved in Demosthenes 35.10–13, "the sole surviving text of an actual Athenian sea-finance agreement" (Cohen 1992, page 42), apart from the aorists that record what has been done, the terms of the agreement are expressed in six finite verbs in the future indicative, four imperatives, and one infinitive.[93]

Clearly, then, in this syntactic environment of contracts and leasing documents in 4th century B.C. Athens, obligations to the πόλις undertaken by ὁ μισθωσάμενος, ὁ πριάμενος, *vel sim.* were often spelled out in the future indicative. It is in this kind of language too that Agyrrhios in 374/3 B.C. chose to outline the contract between the tax-farmers of the *dodekate* ([ο]ἱ πριάμενοι τὴν δωδεκάτην, lines 46–47) on the one hand and ἡ πόλις on the other. Perhaps his own previous experience as a tax-farmer, his familiarity with contracts, finance, and banking, and his involvement with at least one prolific importer of grain to the Peiraieus (see above, page 22) led him to depart from the customary imperatives and infinitives employed by contemporary lawgivers in favor of a more vivid, perhaps even more technical, style. As a result, I suggest that his text of lines 9–36 closely resembles in many respects the form of a contract. We might be tempted to go farther and suggest that it resembled the contract that the tax-farmers of the *dodekate* and the *pentekoste* entered into with the *boule* and the *poletai* after the auction. Specified in the latter document, of course, would have been the name of the πριάμενος, the names of his ἐγγυηταί, and, at a minimum, the number of μερίδες of wheat and barley he had contracted with the πόλις to deliver to the Aiakeion.

Much of the preceding paragraph will probably be rejected by those who minimize the use of written contracts in Classical Athens[94] and by those who would have us believe that most public inscriptions were laboriously cut, sometimes at considerable cost, only to be erected on large stone stelai in prominent locations in Athens where they were seldom read.[95] Can we really believe,

[93] Outside of Athens I cite only one instructive parallel in which the future indicative predominates: the Hellenistic building contracts for the Temple of Zeus Basileus at Lebadeia, *IG* VII 3073–3076; *BCH* 15, 1891, p. 449; 20, 1896, pp. 323–325; *AM* 22, 1897, pp. 179–182 (*SEG* XXXIV 354; XXXVIII 383). The best texts of these inscriptions known to me are in the unpublished dissertation of Turner (1994, pp. 264–361).

[94] For recent statements of this majority view, endorsed by distinguished predecessors such as Pringsheim and Finley, see Bogaert 1968, p. 384; W. V. Harris 1989, pp. 68–70; Thomas 1989, pp. 29, 41–42, 55–58, repeated in Thomas 1992, pp. 89, 149; Cohen 1992, pp. 124–125, 178–179, with earlier bibliography.

[95] Among extreme formulations of this view is Thomas (1989, p. 35): "It is not clear that Athenians actually read inscriptions much"; p. 51: "Even the clause that sometimes occurs at the base of an Athenian inscription describing the purpose of its erection is perhaps ambiguous; a stele is said to be erected 'so that anyone who wishes may see it' (*skopein*)—but not necessarily read it." In fact, this clause does not occur "at the base of an Athenian inscription" and almost never describes "the purpose of its erection." There is, of course, no ambiguity in the formula, as the much more common variant εἰδέναι τῶι βουλομένωι, *IG* I^3 84, line 26, and numerous similar constructions make clear. The appropriate citation to explain epigraphic σκοπεῖν τῶι βουλομένωι is not a medieval "parallel" (Thomas, note 122), but Thucydides 1.22.4, ὅσοι δὲ βουλήσονται . . . τὸ σαφὲς σκοπεῖν. In both cases the infinitive surely indicates close scrutiny, not just some kind of symbolic "seeing." On p. 61, Thomas asserts that "*it is common sense* that the mere presence of documents, even public ones, does not prove that they were read or used" (my italics); also p. 67. This is all toned down a bit in Thomas 1992, pp. 84–88, 139–140. For similar views in the writings of C. W. Hedrick, Jr., see *SEG* XLIV 247. For valuable correctives, see Sickinger 1994a; 1994b; Pritchett 1996, pp. 14–39. I plan to take up the question "Did the Ancient Athenians Read Their Inscriptions?" in a separate publication.

however, that this stele would have been ignored or looked upon merely as a symbol by potential πριάμενοι, by entrepreneurs (of whom there may have been many), by men who could be called upon to act as guarantors, by bankers who might profit from loans financing voyages to Lemnos, Imbros, and Skyros, by Athenian citizens who might be elected as one of the "ten men," by owners of carts or pack animals eager for the business of conveying the public's grain from the Peiraieus to Athens, by owners of ships that might be rented by the πριάμενοι to transport the grain to the Peiraieus, by contractors needed to carry out the alterations to the Aiakeion, and finally by many members of the Athenian *demos* interested in learning details about the origin, the cost, the storage, the quality, the time and place of the sale of their public grain—and possibly also the allocation of the proceeds of that sale to the military fund?

Embodied in a law enacted by a limited number of *nomothetai*, some of this vital information may not have been publicly debated in the Assembly. We know very little about the interaction of the *nomothetai* and the *ekklesia*.[96] Our stele might have been the principal means of communication of the law of the *nomothetai* to the public. It seems likely to me that this inscription was the object of considerable scrutiny, perhaps accompanied by animated discussions on the spot. Moreover, it strains credulity to imagine that the implementation of this complex law, the contractual obligations of the πριάμενοι and the πόλις, the recording of the exact number of μερίδες purchased by so-and-so, and so forth, were all left entirely to what has been called "the oral method."[97] Oral attestation by witnesses was certainly required, of course, in the form of the two guarantors per μερίς, but lines 9–36 of our inscription are themselves, in a sense, a written contract, publicly displayed, merely lacking, among other things, the names of the πριάμενοι, the ἐγγυηταί, and the number of μερίδες. In my view our stele presupposes written συγγραφαί recording these important data and not merely oral attestation by guarantors that the πριάμενοι will live up to the terms of the law as it stood on the stele.

Presumably, this information would also have been recorded by the *boule* and/or the *poletai* after the auction. Their records would have to have been available also in some form to the ten men who had to receive the wheat and barley as it came into the Aiakeion. They had to keep track of how much each πριάμενος had delivered and possibly even when. We know that they kept records of the sale of the public grain because in lines 52–53 they are instructed λογισάσθων ἐν τῶι δήμωι. These records were surely written, and they are unlikely to have been restricted merely to the sale of the public's grain. They probably also included data on how much grain the ten men had received and from whom, for a watchful *demos* was unlikely to have entrusted such potentially profitable property to the memories of its elected representatives for at least three months without requiring some tangible means of keeping track of it.

This is not the place for a full discussion, but the study of our recently discovered inscription suggests that the topic of the form, use, frequency, and the development of written contracts in Athens might repay reinvestigation, preferably with less attention lavished on anthropology and medieval "parallels" and more care expended on a detailed assessment of the epigraphic—as well as the literary—evidence.[98] Van Berchem (1991) has provided an excellent stimulus toward such

[96] See Stroud 1974, pp. 162–163; Hansen 1991, pp. 167–169.

[97] For this term see Thomas 1989, p. 29, followed, unfortunately, by Cohen (1992, pp. 178–179), in an otherwise characteristically astute analysis of the potentially partisan role of witnesses to maritime loan contracts.

[98] For instance such documents as the συγγραφαί for the leasing of the temenos of Kodros, Neleus, and Basile in 418/7 B.C., *IG* I³ 84, or those mentioned in *IG* I³ 402, line 19, which might shed light on the development of public written contracts. While many 5th-century B.C. συγγραφαί seem best interpreted as written specifications to which all citizens must adhere (e.g., *IG* I³ 21 and 78), those on *IG* I³ 84 are clearly directed at a specific individual, ὁ μισθωσάμενος, who will enter into an agreement with the πόλις for the leasing of this temenos. The specific conditions of his contract with the state are specially labeled in lines 29–38 and spelled out in writing on the stone. Like lines 9–36 of Agyrrhios' law, these terms of the contract lack only the name of the μισθωσάμενος and his guarantors or witnesses. When the lease was awarded to him, did this man, following the oral method, merely stand in front of the stele or an altar with his guarantors and swear to carry out the obligations of the lease, or

an investigation. Arguments from etymology can be dangerous, but if written contracts were such a rare and relatively late phenomenon in Classical Athens as some would have us believe,[99] it is peculiar that the Greeks chose to call them συγγραφαί rather than some other word that would more accurately have evoked the "oral method."

LINES 10–14

[κο]|μιεῖ τὸν σῖτον κινδύνωι τῶι ἑαυτõ ὁ π̣[ρ]-
ιάμενος εἰς τὸν Πειραιᾶ καὶ ἀνακομι[ε]-
ῖ εἰς τὸ ἄστυ τὸν σῖτον τέλεσιν τοῖς α[ὑ]|τõ

The buyer of the tax will convey the grain to the Peiraieus at his own risk and he will transport the grain up to the city at his own expense

Here, and in lines 47–48 and 50, Agyrrhios uses the same verb for the shipping of grain from Lemnos, Imbros, and Skyros to the Peiraieus as we have seen in the Eleusinian accounts of 329/8 B.C., lines 288–289, τῆς ἀπαρχῆς . . . ἣν ἐκόμισε Χαιρέστρατος ("of the first-fruits that Chairestratos has conveyed"), above, page 33. It was a standard term for grain: see, e.g., Aristotle, *Ath* 51.4, καὶ τοῦ σίτου τοῦ καταπλέοντος εἰς τὸ σιτικὸν ἐμπόριον τὰ δύο μέρη τοὺς ἐμπόρους ἀναγκάζειν εἰς τὸ ἄστυ κομίζειν ("and of the grain that sails into the grain market, it is incumbent upon the merchants to convey two-thirds to the city"). Note that this regulation to the effect that *emporoi* had to bring two-thirds of each cargo of *sitos* up to the *asty* (while one-third remained in the Peiraieus) did not apply to the *priamenoi* of the *dodekate* in the islands. They had to bring all the tax-grain to the *asty*. For a plausible explanation of the two-thirds rule, see Gauthier 1981, pages 19–28.

To distinguish the first leg of the journey by sea, which was subject to κίνδυνος (see below, pages 49–50), from the final stage, which involved the πριάμενος in τέλη (see below, pages 50–51), the lawgiver shifts to the more graphic compound ἀνακομιεῖ εἰς τὸ ἄστυ τὸν σῖτον in lines 12–13, 18–19, and 19–20. Again, this was standard terminology for bringing commodities—or even merely for walking—inland, up from the harbor at the Peiraieus to the city. See, e.g., Bekker, *Anecdota Graeca* 1.253.26–254.2 = *Agora* XXVIII, page 158, no. 85, ἄνω μὲν ἐν ἄστει, κάτω δὲ τὸ ἐν Πειραιεῖ.

Note the careful distinction drawn throughout between ἄστυ, lines 13, 18–19, and 20, and πόλις, lines 16, 17, 20–21, and 33. The former is a physical place, where the Aiakeion is located. The latter is an entity, a partner to the contract, representing a congeries of unspecified public officials, who act on behalf of the state to provide the Aiakeion with a roof and a door (line 15), to bypass the collection of rent from the πριάμενοι for storage of the grain (line 20), and to exact the grain from a *symmoria* (line 33). *Polis* is not used in this inscription in the physical sense of "the city."

---

were written documents bearing his name and those of his guarantors also employed? We learn from lines 24–25 that, at the very least, his name and those of his guarantors were written up on the wall.

[99] "The written contract first appears in our evidence in the first decade of the fourth century (Isoc. *Trapez.* XVII 20) and more frequently thereafter" (Thomas 1989, p. 41). "Prior to the first decade of the fourth century, written contracts were unknown, and agreements were entirely oral, their contents to be established by oral attestation from witnesses" (Cohen 1992, pp. 178–179). Cf. Gofas 1993, pp. 229–232, with copious citations and bibliography.

For an instructive example of the scarcity of literary references to Greek business letters before the 4th century B.C., now sharply contradicted by at least nine surviving specimens on lead and bronze, see the helpful observations of Millett (1991, pp. 259–260, note 27). The growing numbers of such documents (see *SEG* XLII 1750) should lead to substantial revision of current minimalist notions as to the importance of written documents in ancient Greek business.

For the use of the term κινδύνωι τῶι ἑαυτõ to indicate risk in maritime trade, see Lysias 32.25; [Demosthenes] 52.20. See also, for the grain trade, the Samian Grain Law, *Syll.*^{3} 976, lines 50–52, τὴν δὲ δ[ι]|εγγύησιν ποιείσθωσαν οἱ ἄνδρες οἱ χειροτονηθέντες ἐπ[ὶ]| τοῦ σίτου κινδύν[ω]ι τῶι ἑαυτῶν ("let the men who have been elected to look after the grain provide security at their own risk"), and for risk in mining, Xenophon, *Poroi* 4.28–29, κίνδυνος δὲ μέγας τῷ καινοτομοῦντι . . . εἰς τοῦτον οὖν τὸν κίνδυνον οὐ μάλα πως ἐθέλουσιν οἱ νῦν ἰέναι ("there is a severe risk for the one who makes a new cutting . . . and so men today are not at all willing to encounter this risk"). For κίνδυνος in a contract, see *PLond.* II, 301, page 256 (= Mitteis, *Chrest.* 340, lines 13–14, A.D. 138–161).

The dangers faced by a πριάμενος in transporting the people's grain from Lemnos, Imbros, and Skyros to the Peiraieus included storms at sea, running aground or shipwreck through faulty navigation, attack and looting by pirates or by enemy warships, seizure and diversion (τὸ κατάγειν) to a foreign port where the cargo might be confiscated, spoilage of the wheat and barley in poor storage facilities on board, and other hazards, some of them perhaps unforeseen. We lack the data to determine the frequency of shipwreck in ancient Greek maritime commerce. Cohen (1992, pages 141–143) emphasizes that modern scholars have tended to exaggerate such dangers: "The loss of a ship in fourth-century Hellas seems akin to the crash of an aircraft at the present time: an occurrence whose infrequency results in enormous public attention." By inserting the phrase κινδύνωι τῶι ἑαυτõ, the lawgiver seems explicitly to have removed the *polis* from even partial responsibility for any such losses. Even though the ships in question in some sense were engaged in a state-sanctioned activity, no guarantee of protection in the form, say, of an Athenian naval escort is proffered.

This is noteworthy, especially in 374/3 B.C., for the Athenians had recently become painfully aware of the perils of leaving grain ships unattended, in the summer of 376 B.C. on the eve of the battle of Naxos. Many grain transports bound for the Peiraieus at that time could not complete the last stage of their voyage from the southern tip of Euboia around Sounion and into the Saronic Gulf because of the arrival of a Spartan fleet of sixty triremes that cruised around Aigina, Keos, and Andros. The precariousness of their food supply must have been clear to many Athenians (Xenophon, *HG* 5.4.60–61; Diodoros 15.34.3–35.2). On this campaign see below, pages 119–120.

Athenian concern for the safety of the ships bringing wheat and barley from Lemnos, Imbros, and Skyros, however, seems not to have been intense enough to win recognition from the *nomothetai* in this particular law. On other occasions we learn of warships escorting convoys of grain transports bound for Athens,[100] but it appears that individual generals normally received such assignments through specific *ad hoc* decrees of the assembly.[101] There may have been some reluctance to build this kind of provision into something as permanent as a *nomos*.

Providing a naval escort of triremes for the ships used by the πριάμενοι to convey grain from the three islands to the Peiraieus might also have been fairly complex and costly. Given what appears to have been the relative freedom enjoyed by the πριάμενοι in collecting the tax and the "private enterprise" character of the whole process, it is possible that all of them might not have readily agreed to sail from the same ports at the same time in orderly convoys. The cargoes of some of these ships may even have contained commodities other than grain produced by the *dodekate* and the *pentekoste*. Each *meris* was relatively small. Moreover, cargo ships and triremes traveled at different speeds and on very different schedules.[102] Merchants, whose profits could depend on getting to port promptly, might have balked at having to tie up or anchor each night while the

[100] For Athenian naval convoys protecting grain ships in the 4th century B.C., see, e.g., [Demosthenes] 50.6, 17, 20–22 (362 B.C.); *IG* II^2 1629, lines 217–232; Gernet 1909, pp. 353–364; Velissaropoulos 1980, pp. 135–136; Garnsey 1988, pp. 142–144; Pritchett 1991, pp. 330–334, 338–341.

[101] This was certainly the case in 362 B.C.; cf. [Demosthenes] 50.6.

[102] Still worth reading on this point is the classic paper of Gomme (1933b, pp. 16–24). See also Casson 1971, pp. 281–296; 1991, pp. 97–115; and Wallinga 1993, pp. 33–65.

trireme crews ate and rested on shore. Maritime risk was probably nothing new to the πριάμενοι of Agyrrhios' law. It is not surprising, then, that they had to convey the grain to the Peiraieus with the threat of loss entirely on their side. For helpful discussions of the many risks that merchant ships had to face, see Isager and Hansen 1975, pages 55–57; Velissaropoulos 1980, pages 127–165, 328–335; De Ste. Croix 1972, pages 47–48 and appendix VIII; Garnsey 1988, pages 142–144; McKechnie 1989, pages 101–141 (esp. 127–128); Adams 1989.

Our law does not state how a defaulting πριάμενος made restitution. If he lost the state's grain at sea through shipwreck, negligence, a piratical raid, or being forced to sail to a foreign port to unload his cargo, what were his legal obligations? Private maritime contracts sometimes made allowances for some of these contingencies by adjusting the scale of repayment required of a borrower.[103] There may have been similar catastrophe-clauses in the Athenian laws governing the collection of taxes, the τελωνικοὶ νόμοι; see above, pages 28–29. Interpreted strictly, however, Agyrrhios' phrase κινδύνωι τῶι ἑαυτõ probably means that doing business with the *polis* was a tougher proposition. If failure to deliver the state's grain on time in the specified condition was entirely the fault of the πριάμενος, we might guess that he and/or his guarantors had to make restitution either by bringing to the Aiakeion an amount of wheat and barley equivalent to the number of *merides* for which he had contracted or by paying in cash a sum calculated on the basis of the price of the grain as set by the *demos* in the manner prescribed in lines 44–46. It would also be in keeping with some other contracts between the *polis* and private contractors, if default brought with it a fine in addition to restitution.[104] Instructive for breach of contract by farmers of taxes collected in cash is Demosthenes 24.122, οὗτος γὰρ . . . τοῖς μὲν τὰ τέλη ὠνουμένοις ἔγραψε τὰς τιμωρίας εἶναι, εἰ μὴ καταβάλοιεν τὰ χρήματα, κατὰ τοὺς νόμους τοὺς προτέρους, ἐν οἷς καὶ ὁ δεσμὸς καὶ ἡ διπλασία γέγραπται ἀνθρώποις, οἳ διὰ τὸ ζημιοῦσθαι ἐπὶ τῇ ὠνῇ ἄκοντες ἔμελλον τὴν πόλιν ἀδικήσειν ("for this man wrote that for the purchasers of [the right to collect] the taxes, the penalties for failure to meet their cash payments are to be in accord with the previous laws that record both imprisonment and a double fine for men who, because of the loss they suffer over their contract, could commit an injustice against the *polis* unintentionally").

Transportation of the grain from the Peiraieus to the Aiakeion in the *asty* was also the responsibility of the πριάμενος. He had to absorb all the expenses of this operation, τέλεσιν τοῖς α̣[ὐ]τõ, lines 13–14. This terminology is standard for setting up a stele, cf., e.g., *IG* I^3 62, lines 10–11; 63, line 7; 66, line 22. Closer to our context are expenses concerned with leasing such as *IG* I^3 84, lines 13–14, τὸν μισθοσάμενον ἕρχσαι τὸ *h*ιε[ρ]ὸν τοῖς ἑαυτõ τέλεσιν ("the lessee is to fence the sanctuary at his own expense"); Pleket 1964, no. 43, lines 11–14, ἐνοι[κ]οδομεῖσθαι δὲ Θρασ[ύ]βουλον ἐάν τι βούληται τοῖς αὑτοῦ τέλ[ε]σιν ("let Thrasyboulos construct something therein, if he wishes, at his own expense"). Since τέλος is also the word for a tax, the πριάμενοι presumably had to pay any such levies that applied to their unloading the grain at the harbor and transporting it to the *asty*. See, e.g., *IG* I^3 63, lines 14–17: [τὸς δὲ βολομένος α]||ὐτõν ἄγεν καὶ σῖτον κ[ατὰ τὰ φσεφίσματα τὰ ἐφσε]||φισμένα τõι δέμοι κ[αὶ ἐσεμπορεύεσθαι τελõντ]||ας τὰ τέλε, *h*ὰ ἂν φσε[φίσεται ὁ δε͂μος ὁ Ἀθεναίον] ("let those among them who wish also carry grain in accord with the decrees that the *demos* has decreed and let them come in to trade, provided that they pay whatever taxes the *demos* of the Athenians has decreed"). See also the tantalizing reference in Hesychios, *s.v.* διαπύλιον· τέλος τι παρ' Ἀθηναίοις οὕτω ἐκαλεῖτο ("*diapylion:* the Athenians gave this name to a tax"), with the discussion by Pritchett (1980, pages 183–184).

In the present context, the πριάμενος probably had to cover the cost of maintaining and/or renting carts or pack animals, and their attendants, for conveying the wheat and barley to the

[103] See the useful discussion in Cohen 1992, pp. 53–58.

[104] For fines of this nature in leasing documents, see, e.g., Behrend 1970, pp. 127–136.

Aiakeion. Pack animals, the σιτοφόροι ἡμίονες of Herodotus 3.153.1, could probably cover the distance between the Peiraieus and the Aiakeion in less time than carts and could follow a more direct, i.e., steeper, route up into the city. Carts, though slower and favoring a more level approach, probably had a greater capacity than sacks or the saddlebags of pack animals. For speculation on the capacity of grain in the latter, see Gauthier 1981, pages 22–23, and the anecdote from Zenobios 1.74, discussed by Pritchett (1980, page 184). For grain carts, see Thucydides 6.7.1; Xenophon, *Kyropaideia* 2.4.18; Raepsaet 1988, with helpful earlier bibliography. In view of Hammond's (1983) demonstration that carts were much more frequently used than pack animals in military transport, we might expect them to have been preferred in the domestic context of our law, especially since there was a wagon road conveniently available. See also Pritchett 1980, pages 181–196. Gofas (1993, pages 221–222) discusses transportation of grain in amphoras, in sacks, and in bulk.

For speculation about the route(s) followed by the tax-farmers in conveying grain to the Aiakeion, see below, pages 104–107.

LINES 14–15

κατανήσει τὸν σῖτον εἰς τὸ Αἰά̣[κ]|ειον

He will heap up the grain in the Aiakeion

When he reached his destination and unloaded his cart or pack animals, the *priamenos* had to heap up his grain in the Aiakeion. For this fairly rare verb, see what Herodotus says about Datis at Delos, 6.97.2, λιβανωτοῦ τριηκόσια τάλαντα κατανήσας ἐπὶ τοῦ βωμοῦ ἐθυμίησε ("heaping up three hundred talents of incense on the altar, he burned them"). The simplex νέω (νηέω), sometimes glossed as σωρεύω, is used of the earth of a funeral mound (Thucydides 2.52.4), bread (Xenophon, *Anabasis* 5.4.27), amphoras (Aristophanes, *Clouds* 1203), clothing (*IG* II^2 1522, line 23), and wood for a fire (Homer, *Odyssey* 19.64).

It is time to define the term ὁ πριάμενος, which is the subject of the three verbs κομιεῖ, ἀνακομιεῖ, and κατανήσει in lines 10–14, ἀποστήσει in lines 16 and 21–22, οὐ θήσει in line 27, and καταστήσ[ε]ι in lines 29–30. In the plural these men appear as one of the double objects of ἐνοίκιον οὐ πράξει [ἡ π]όλις, lines 20–21, and as the subject of τὸν σ⟨ῖ⟩τον . . . κομισάντων in lines 46–48. I believe that the same persons are intended in all nine passages and that the shift from singular to plural and vice versa is not significant. Who, then, is "the buyer" and what is he buying?

Possibilities seem to be limited to grain and to the *dodekate* and the *pentekoste*. Are we in the first instance to understand τὸν σῖτον as the unexpressed object of ὁ πριάμενος throughout? Is the *polis* intervening somehow in the buying of wheat and barley from the islands to the extent of extracting from "the buyer" $8\frac{1}{3}$% of the volume sold, or $8\frac{1}{3}$% of the cash value of the grain sold? Is the *polis* instructing those who buy the grain to convey it, or a portion of it, to the Aiakeion for storage and resale later?

One obstacle in the way of this kind of reconstruction is the then inexplicable provision in line 27, προκαταβολὴν οὐ θήσει ὁ πριάμενος (τὸν σῖτον), "the buyer [of the grain] will not pay a down payment." Why would anyone have expected him to do so and why is the buyer (of the grain) in the next line assessed ἐπώνια καὶ κηρύκεια and required to produce guarantors?

Fortunately, we do not need to prolong this unlikely scenario, for in lines 46–48 the words for both "grain" and the *dodekate* appear together with the πριάμενοι in the formulation τὸν δὲ σ⟨ῖ⟩τον [ο]ἱ πριάμενοι τὴν δωδεκάτην κομισάντων, which clearly shows that it is the *dodekate* that the buyers buy, and not the grain. I suggest, therefore, that throughout this inscription we should understand τὴν δωδεκάτην as the unexpressed object of ὁ πριάμενος. This means, of course, that everywhere "the buyers" remain the same people, i.e., the buyers of (the right to collect) the $8\frac{1}{3}$% tax on the grain in the islands. This is in keeping with lines 6–7, which explicitly state that it is the tax that is being sold, τὴν δωδεκάτην πωλ[εῖ]ν. The lawgiver then closely follows this instruction with a definition of this tax in terms of *merides* of wheat and barley. Then in lines 10–14 the buyer (of the tax) receives instructions about the transportation of this *sitos* to Athens. Since the buyer (of the tax) is assessed κηρύκεια in line 28, he must have gone through the process of an auction. I see then no cogent objection to interpreting "the buyer" throughout the inscription as one who, after bidding successfully at an auction, has purchased the right to collect the *dodekate* on the grain in Lemnos, Imbros, and Skyros.

There is potential for confusion in this interpretation, for in the four passages where the verb πωλέω appears in this inscription, it does not always govern objects which seem to modern readers to be in the same category. Given the conventions of Athenian tax terminology, however, this is hardly surprising. In lines 6–7, τὴν δωδεκάτην πωλ[εῖ]ν means "to sell (the right to collect) the $8\frac{1}{3}$% tax," whereas the ten men elected by the *ekklesia* are told (lines 40–41) τὸν σῖτον ... πωλόντων ἐν τῆι ἀγ[ορ]ᾶι, "let them sell the grain in the Agora." In further instructions to the ten men, πωλε̃ν in lines 42 and 46 means again "to sell the grain," and Agyrrhios conveyed this same meaning in lines 51–52 with the variant, ἐπειδὰν δὲ ἀποδῶνται οἱ αἱρεθέντες τὸν σῖτον, "whenever the elected men sell the grain." Unlike ὁ πριάμενος, which in my view consistently governs the same unexpressed object (τὴν δωδεκάτην) throughout, πωλεῖν appears to vary in meaning as its expressed object changes from τὴν δωδεκάτην to τὸν σῖτον. On this interpretation, the buyers of the tax, who are the lawgiver's primary concern, are mentioned in nine passages of this inscription, whereas those who will buy the grain from the ten men in the Agora (lines 41–46) are never mentioned.

For ὁ πριάμενος as the one who purchases the right to collect a tax, see, among many examples, Andokides 1.73, ὠνὰς πριάμενοι ἐκ τοῦ δημοσίου μὴ κατέβαλον τὰ χρήματα ("although they purchased [the right to collect] taxes from the state, they did not make their cash payments"); 92–93, Κηφίσιος μὲν οὑτοσὶ πριάμενος ὠνὴν ἐκ τοῦ δημοσίου, τὰς ἐκ ταύτης ἐπικαρπίας τῶν ἐν τῇ γῇ γεωργούντων ἐνενήκοντα μνᾶς ἐκλέξας, οὐ κατέβαλε τῇ πόλει καὶ ἔφυγεν· εἰ γὰρ ἦλθεν, ἐδέδετ' ἂν ἐν τῷ ξύλῳ. ὁ γὰρ νόμος οὕτως εἶχε· κυρίαν εἶναι τὴν βουλήν, ὃς ἂν πριάμενος τέλος μὴ καταβάλῃ, δεῖν εἰς τὸ ξύλον ("Now this fellow Kephisios having purchased [the right to collect] a tax from the state and after collecting the revenues it produced from those who were farming the land to the total of ninety minai, failed to make payment to the *polis* and fled. For if he had returned, he would have been confined in the stocks. For the law provided that the *boule* is authorized to confine in the stocks any tax-purchaser who does not make his payments"); Demosthenes 24.144, ἐάν τις ἐπὶ προδοσίᾳ τῆς πόλεως ἢ ἐπὶ καταλύσει τοῦ δήμου συνιὼν ἁλῷ ἢ τέλος τι πριάμενος ἢ ἐγγυησάμενος ἢ ἐκλέγων μὴ καταβάλλῃ ("if someone is convicted of knowingly betraying the *polis* or subverting the *demos* or anyone who has bought [the right to collect] a tax or serve as a guarantor or collector and does not make his payment"), cf. 24.40–41, 122; Pollux 9.33–34, τὸ μέντοι τὰ τέλη παραλαβεῖν εἴποις ἂν πρίασθαι ("you might say that to assume taxes is to buy them"); Kos, *AM* 16, 1891, page 406, line 4, τοὶ πριάμενοι τὰν ὠνὰν σίτου ("the purchaser of the [right to collect the tax] on grain"), with J. Toepffer on pages 415–417; *Syll.*[3] 1000.

The participle is also found in contexts of leasing where someone "buys" the right to lease a property or buys a contract to make repairs on a building, *vel sim.*; see *IG* II^2 1176, *passim*, οἱ πριάμενοι τὸ θέατρον in the Peiraieus; see new text in *SEG* XXXIII 143; Gauthier 1976, pages 148–149. See the discussion of the terminology of buying and selling in the context of leasing and taxation by Rosivach (1992, pages 57–59); see also Lambert 1997, pages 265–268.

LINES 15–16

στέγον δὲ καὶ τεθυρωμένον παρέ[ξ]|ει τὸ Αἰάκειον ἡ πόλις

The state (πόλις) *will make available the Aiakeion in watertight condition and provided with a door.*

Στέγον is the neuter active participle of στέγω, "to keep out water," "to be impervious," cf. Aeschylus, *Seven* 216, 234, 797, of ships and towers; Thucydides 2.94.3, of ships; τὴν οἰκίαν . . . στέγουσαν, *IG* II^2 2498, lines 22–23; XII.5.568, line 12. It is often, as here, combined with τεθυρωμένον, "provided with a door," as in the Herakleia Tablets, *IG* XIV 645, I, lines 141–142, Uguzzoni and Ghinatti 1968 *ad loc.*, ταῦτα δὲ παρεξόντι οἰκοδομημένα καὶ στεγόμενα καὶ τεθυρωμένα ἐν τοῖς χρόνοις ἐν οἷς καὶ τὰ δένδρεα δεῖ πεφυτευκῆμεν ("these structures they will provide in watertight condition and with doors in the time in which it is also necessary to plant the trees"), cf. *IG* II^2 1046, lines 14–18.

Without committing myself at this point to inferences about the nature and appearance of the Aiakeion, I suggest, nevertheless, that the future tense παρέξει and the context imply that the *polis* may have had some work to do on this sanctuary before it would be ready to serve as a public granary. For details about constructing granaries, see the edition of Philon of Byzantion Book "V" 86.6–88.30, in Garlan 1974, pages 300–303.

For speculation on the identification of the Aiakeion, which I place in the southwest corner of the Agora, see below, Chapter III.

LINES 16–19

ἀποστήσε[ι τ]-
ὸν σῖτον τῆι πόληι τριάκοντα ἡμερῶν [ὁ]
πριάμενος, ἐπειδὰν ἀνακομίσηι εἰς [ἄσ]-
τυ τέλεσι τοῖς αὑτô

The buyer [of the tax] is to weigh out the grain for the state within thirty days, whenever he has conveyed it up to the city at his own expense

The verb ἀφίστημι appears in lines 21–22 and 24–25, where the process it denotes is described in detail, and in lines 40–41 the newly elected "ten men" who take care of the grain also ἀποστησάμενοι τὸν σῖτον κ[α]τὰ τὰ γεγραμμένα. These men represent τῆι πόληι of line 17. I take ἀφίστημι in all these passages as transitive in the sense of "weigh out," as in *IG* I^3 52A, lines 20–21 (the 5th-century B.C. Kallias decree), ἀπαριθμεσάσθον καὶ ἀποστεσάσθον τὰ χρέματα ἐναντίον τε̃ς βολ[ε̃]ς ἐμ πόλει ("let them count and weigh out the money in the presence of the *boule* on the Acropolis"); Xenophon, *Symposion* 2.20, οὕτω τὰ σκέλη τοῖς ὤμοις φαίνῃ ἰσοφόρα ἔχειν ὥστε δοκεῖς ἐμοί, κἂν εἰ τοῖς ἀγορανόμοις ἀφισταίης ὥσπερ ἄρτους τὰ κάτω πρὸς τὰ ἄνω, ἀζήμιος ἂν γενέσθαι ("your legs seem to be of such equal weight with your shoulders that you look to me as if you would not incur a fine were you to weigh out your lower parts against your upper parts for the *agoranomoi* just like loaves of bread"), see the excellent discussion of this passage by Ampolo (1986); *IG* XII.5.1.647, line 15, καὶ ἀφίστασθαι τὰ κρέα ("weigh out the meat"), Keos ca. 3rd century B.C.

The interval or grace period of thirty days, a standard length,[105] probably gave the ten men and the πριάμενοι enough time to complete the weighing and quality control in the face of inevitable delays such as festival days, disputes over the weight and quality of the grain, large amounts of produce still awaiting processing, and so forth.

[105] For the limit of thirty days in legal contexts and for paying money, see MacDowell 1990, pp. 266–267.

LINES 20–21

ἐνοίκιον οὐ πράξει [ἡ π]|όλις τοὺς πριαμένους

The state will not exact rent from the buyers of the tax

Since the Aiakeion is a public granary for the storage of the people's wheat and barley, the πόλις will not exact rent or a storage fee (ἐνοίκιον) from the πριάμενοι for the use of its facilities during the interval between the arrival of the grain in the Aiakeion and its final turning over to the ten men. The repairs to the Aiakeion and this waiver of the rent both indicate that this temenos was a public building at the disposal of the *polis*. The inclusion of this provision in the law may suggest that in the private sector, grain merchants were normally charged a storage fee for their produce from the time it arrived in the grain market until they sold it.[106] That this practice was not restricted to the private sector is made clear by a passage in the ever-helpful accounts of the Eleusinian *epistatai* for 329/8 B.C., *IG* II^2 1672. In lines 296–300 the *epistatai* record a storage fee, ἐνοίκιον, of one drachma, one obol that they had to pay on the *aparche* from Imbros when the general Chairestratos brought it in late, after the sacrifices were over. They did not have to pay any rent or storage fee on all of the other grain of the *aparchai* from the Attic tribes and Athenian possessions abroad listed in lines 263–295, which presumably arrived according to schedule. Gauthier 1989, pages 101–107 has an excellent discussion of ἐνοίκιον of commercial buildings with copious, apt parallels.

For the use of the verb πράττω here, see below, pages 66–67.

LINES 21–25

τοὺς πυροὺς ἀ̣[πο]-
στήσει ὁ πριάμενος ἕλκοντας πέντε ἑ̣[κ]-
τέ⟨α⟩ς τὸ τάλαντον, τὰς δὲ κρι⟨θ⟩ὰς ἑλκο[ύσ]-
⟨α⟩ς τὸν μέδιμνον τάλαντον ξηρὰς ἀπο̣σ[τ]-
ήσει καθαρὰς αἰρῶν

The buyer of the tax will weigh out the wheat at a weight of a talent for five sixths and the barley at a weight of a talent per measure (medimnos) *dry and free from darnel*

Here is potentially valuable new evidence for the procedure of measuring and weighing wheat and barley in 4th-century B.C. Athens. It is clear from the instructions in lines 14–15, κατανήσει τὸν σῖτον εἰς τὸ Αἰά̣[κ]ειον, and 16–17, ἀποστήσε[ι τ]ὸν σῖτον τῆι πόληι, that the πριάμενος will weigh his wheat and barley at or in the Aiakeion. Although in lines 21–27, it is the individual πριάμενος who is instructed to weigh his grain (ἀ̣[πο]στήσει ὁ πριάμενος [21–22] . . . ἀπο̣στήσει [24] . . . σηκώσας [26]), later, in specifying the duties of the ten newly elected magistrates, Agyrrhios clearly indicates that these officials are to oversee this operation, οὗτοι δὲ ἀποστησάμενοι τὸν σῖτο̣ν κ[α]τὰ τὰ γεγραμμένα, lines 40–41. Presumably these magistrates also at this time examined the barley to make sure that it is dry and free of darnel seeds (ξηρὰς . . . καθαρὰς αἰρῶν, lines 24–25).

In this section of the law, the weights come first: wheat—five *hekteis*[107] weigh one talent; barley—one *medimnos* weighs one talent. To my knowledge, this is the earliest evidence for weight/volume ratios of wheat and barley from ancient Greece. As expected, wheat in our law is heavier than barley, when equal measures are weighed. The weight ratio of wheat to barley of 6 : 5 accords with that of 60 : 48 in Egypt restored in *POxy* 3455 by the editor, J. C. Shelton, and with some modern figures of one bushel of wheat = sixty pounds, one of barley = fifty pounds.[108]

[106] For a charge of ἐνοικείου [θησα]υ̣ρ̣οῦ for wheat, see *PTeb* 520 and 339, lines 17–19, with helpful parallels.

[107] M. Lejeune (1991) has recently clarified the distinction between ἕκτη (a monetary unit and a liquid measure) and ἑκτεύς (a dry measure mainly for wheat and barley).

[108] For the relative weights of wheat and barley, see Pliny, *NH* 18.62; Pritchett 1956a, p. 193; Foxhall and Forbes 1982, pp. 42–44. In general see Jardé 1925, p. 32:

Our weight/volume data are valuable since "in antiquity amounts of grain were normally expressed in units of volume, whereas today grain is handled in units of weight" (Foxhall and Forbes 1982, page 42); cf. Jardé 1925, page 31. In his law, however, Agyrrhios was not content to specify only the number of *medimnoi* of wheat and barley that the πριάμενοι would bring up to the Aiakeion. The *polis* also required that each measure of its grain weigh a prescribed amount. Today, and presumably in antiquity, the weight of wheat and barley per unit of volume can fluctuate considerably owing to variables such as "species and variety of grain, time of year sown, soil conditions, amount and type of fertilizer used, whether grain is hulled or naked, and the amount of foreign bodies, weed seeds, etc. present in a sample" (Foxhall and Forbes 1982, page 42). Local factors such as the geographic origin of the grain may also have played a role.[109] We must be careful, therefore, not to infer that the weight/volume ratios in lines 21–27 represent some kind of standard figures either for Athens or for Greece in general. Agyrrhios knew that this grain was coming from Lemnos, Imbros, and Skyros, and it is possible that he set the weight requirements accordingly.[110]

Nevertheless, our law now brings fresh evidence to bear on the question of the weight of ancient wheat and barley and on estimates advanced by modern scholars. Difficulties of metrological conversion of ancient figures to modern numbers for wheat and barley are notorious. In addition to all the variables mentioned above, we also lack precise, reliable evidence for the capacity of the Attic *medimnos* and the weight of the Attic talent in the first quarter of the 4th century B.C. Lang has well laid out the problems in *Agora* X, pages 2–23, 34–36, 39–48.

If, merely as an estimate, however, we take her (4th century B.C.?) Attic weight standard of a stater of ca. 105 drachmai, or 882–924 gr., and reckon 30 staters to a talent, then the weight of the latter will be ca. 26,460–27,720 gr., or 26.46–27.72 kg. This will be roughly the weight of one *medimnos* of barley. To translate this weight approximately into a measure of capacity, we might, e.g., adopt Lang's figure of 52.416 cc. for a *medimnos* of barley. Five *hekteis* of the heavier wheat, i.e., 43.68 cc., will then weigh the same amount or 1 talent = 26.46–27.72 kg. Or, one *medimnos* of wheat weighs roughly 30.86–32.34 kg.

These weights of ca. 26.46–27.72 kg. for a *medimnos* of barley and ca. 30.86–32.34 kg. for a *medimnos* of wheat from the islands in 374/3 B.C. are considerably lighter than many modern estimates based upon the weight of modern grains.[111] Is this a symptom of our inexact knowledge of the Attic system of metrology at this time, or another warning that modern "parallels" provide a hazardous basis for reconstructing the nature of ancient agricultural products? At any rate, the figures of one talent for five *hekteis* of wheat and one talent for one *medimnos* of barley from Lemnos, Imbros, and Skyros in 374/3 B.C. must now take their place in future calculations of weight/volume ratios of grain in antiquity.

Another factor that can significantly affect the weight of a given unit of grain is, of course, dampness. This might especially have been a factor in grain transported by ships. In lines 23–24,

"les Anciens nous instruisent mal du poids des céréales ... Nous n'avons dans les textes anciens aucune donnée sur le poids de l'orge." This needs now to be qualified also for Egypt in late antiquity; see Rathbone 1983.

[109] See the wide variety in the weights of contemporary wheat from different parts of the Mediterranean recorded by Pliny (*NH* 18.66–70), selectively summarized in Jasny 1944, pp. 63–65, 82–87; Moritz 1958, pp. 168–209; excellent discussion in Rickman 1980, pp. 153, 239–240; Foxhall and Forbes 1982, p. 43; see also Jardé 1925, pp. 31–32; Sallares 1991, p. 488, note 160.

[110] For the possibility of a difference in quality between the grains of Attica and those grown in the three islands, see Sallares 1991, p. 331. Pollux 7.152 (= *PCG* III.2, Aristophanes F 389) noted that the harvest (καρπός) on Lemnos was early. Theophrastos, *CP* 4.9.6, characterizes the wheat of Lemnos as relatively heavy, but I do not believe that this passage substantiates the claim of Figueira (1986, p. 158): "since large consignments of grain could not be weighed, the weight of grain was estimated from volume by knowing its place of origin (Theophr., *CP* 4.9.5)."

[111] Jardé estimates weights of 34.07 kg. for a *medimnos* of barley and 41.95 kg. for wheat. Garnsey (1985, p. 72) reckons 33.75 kg. barley and 42.73 kg. wheat. Cf. also Sallares 1991, p. 79. Sallares 1991, p. 486, note 148: "the mean grain weight of English barley has increased by 15 percent in the course of the twentieth century A.D."—a cautionary tale?

τὰς δὲ κρι⟨θ⟩ὰς . . . ξηράς, our law anticipates any inclination on the part of the πριάμενοι to cheat the *polis* by piling up and weighing out heavier, damp barley. Cf. the *diagramma* of Philip V from Chalkis, lines 15–16, καὶ τὸμ μὲν σῖτον ἀναγέτωσαν ἀπὸ τῆς νέας προσόδου ἄβροχον ("let them bring in the grain from the new harvest in dry condition"): Welles 1938, page 252.[112]

The ten men also exercise quality control over the state's barley by determining that it is free of darnel seed, καθαρὰς αἰρῶν, line 25. One device for removing darnel from harvested grain was the darnel sieve, κόσκινον αἰρόπινον, attested as early as Aristophanes' (undated) play Σκηνὰς Καταλαμβάνουσαι.[113] For a thorough discussion of this weed and its effects on barley, see Sallares 1991, pages 338–340, 482, note 113, 487, note 154. For other examples of quality control of grain, see the Herakleia Tablets, *IG* XIV 645, I, line 103 (Uguzzoni and Ghinatti 1968 *ad loc.)*, κριθᾶς κοθαρᾶς δοκίμας; below, page 59; *IG* V.2.514, from Lykosoura, line 15, ὀλοαῖς αἰρολογημέναις; and numerous examples on papyri, from which I select only two: πυρὸς καθαρὸς κεκοσκινευμένος, *PStrassb* 563, with Hauben 1975, pages 289–291; and πυροῦ καθαροῦ ἀδόλου κεκοσκινευμένου ("wheat that is clean, pure, sifted"), *SB* V.8754, lines 10–12. See also the discussion of κα(θαρός) by Dillery (1996, page 174, no. 781). Cf. what Aristotle says of the Athenian *agoranomoi* in *Ath* 51.1, τούτοις δὲ ὑπὸ τῶν νόμων προστέτακται τῶν ὠνίων ἐπιμελεῖσθαι πάντων, ὅπως καθαρὰ καὶ ἀκίβδηλα πωλήσεται ("these men have been assigned by the laws the task of overseeing the sales of all commodities so that they will be sold clean and unadulterated"). For quality control in Roman Egypt, see Rickman 1980, pages 121–122 and for 6th-century Byzantion, Gofas 1993, pages 243–248.

LINES 25–27

τὸ σ⟨ή⟩κωμα ἐπὶ ΤΗ[. .]|ΩΝ<Η>Ι σηκώσας, καθάπερ οἱ ἄλλοι ἔμ[π]ορ[ο]|ι

Here the law even specifies the method of weighing and measuring the wheat and barley to be followed by the πριάμενος, which lines 40–41 tell us was carried out under the supervision of the ten men. Even though in two cases the mason failed to carve the horizontal stroke of eta, the readings here seem not to be in doubt. The missing strokes may well have been added in paint.[114] There remains, however, the problem of the restoration of ΤΗ[. .]|ΩΝ<Η>Ι. It is intimately related to the interpretation of τὸ σήκωμα and the verb σηκόω.

Whatever procedure the lawgiver envisaged, clearly it had to be in accord with recognized and accepted practice followed by "the other emporoi." The *priamenos* must duplicate a procedure

[112] For the effect of dampness on the weight of grain, see Jardé 1925, p. 41; Garnsey 1988, p. 48; Sallares 1991, p. 344; Gallant 1991, pp. 96–98, 179–181. For precautions against dampness in storing grain, see the edition of Philon of Byzantion Book "V" 86.6–88.30, in Garlan 1974, pp. 300–303. There is no mention in our law of the practice of mixing earth with stored grain to reduce moisture, as outlined by Theophrastos (*HP* 8.11.7) and prescribed in lines 16–18 of the *diagramma* of Philip V.

[113] I owe the correct interpretation of ΚΑΘΑΡΑΣΑΙ-ΡΩΝ here to the acumen of M. H. Jameson, who aptly adduces Theophrastos, *HP* 8.4.6, speaking of wheats, εἰσὶ δὲ καὶ οἱ μὲν καθαροὶ αἰρῶν . . . καὶ μάλιστα ὁ Ἀκραγαντῖνος οὐκ αἰρώδης ("there are some which are also free of darnel . . . and the wheat from Akragas is especially darnel free"); cf. 8.7.1. See also Gallant 1991, p. 50. I have considered and rejected καθαράς, αἶρων τὸ σ⟨ή⟩κωμα. For the darnel sieve, see *PCG* III.2, Aristophanes F 497 (F 480 Koch) from Pollux 10.114. Cf. Bekker, *Anecdota Graeca* 1.22.11, αἰρόπινον· κόσκινον. πίνος γάρ ἐστιν ὁ ῥύπος . . . αἰρόπινον οὖν τὸ κόσκινον τὸ τὰς αἶρας, ὅπερ ἐστὶ ῥύπος τῶν πυρῶν, καθαίρων ("darnel sieve: for impurity is a kind of dirt . . . and so the sieve that cleans out the darnel, which is an impurity in wheat, is a darnel sieve"); Hesychios, *s.v.* αἰρόπινον . . . κόσκινον, ἐν ᾧ πυροὶ σήθονται ("darnel sieve . . . a sifter in which wheat is sifted"). For grain sieves, see Moritz 1958, pp. 157–167; Amyx 1958, pp. 259–261.

[114] Omission of these strokes could conceivably have arisen from the fact that there are four etas fairly close together in the five words before καθάπερ, which could have induced the mason to cut all the verticals first. I suggested above, p. 13, that similar concentrations of alphas in other lines may have led the cutter to omit the crossbar.

that was presumably well enough known that Agyrrhios did not have to repeat all of its provisions here. Was it simply a method based upon long-established custom among merchants, or one that the *polis* had imposed through written regulations? The inclusion of this brief clause in our law specifying how the grain was to be weighed favors the second alternative. We might expect that rules governing methods of weighing and measuring formed part of the Athenian ἐμπορικοὶ νόμοι: [Demosthenes] 35.3, κατὰ τοὺς ἐμπορικοὺς νόμους; Hesychios, *s.v.* ἀστικοὶ νόμοι· οἱ κατὰ τὴν Ἀθηναίων πόλιν, ἦσαν γὰρ καὶ ἐμπορικοί ("city laws: those found in the *polis* of the Athenians, for these were also commercial laws"). For discussion of the content of these laws with helpful earlier bibliography, see Velissaropoulos 1980, pages 235–245; Bresson 1993, pages 165–171; Gofas 1993, pages 167–195. The activity of *emporoi* was being monitored at least as early as 375/4 B.C. by the ἐπιμεληταὶ τοῦ ἐμπορίου.[115] In our law, however, it is the ten men, newly elected (lines 40–41), who are to supervise this process. Under their scrutiny the πριάμενος is to use the *sekoma*, τὸ σ⟨ή⟩κωμα . . . σηκώσας.

Examples of τὸ σήκωμα meaning a weight in Attic inscriptions include the notations painted on the necks of amphoras published in *Agora* XXI, Hb4; He5; He22 (with M. Lang's good notes), and *IG* II^2 1013 + *SEG* XXIV 147–148 (see below, page 58). For this sense see also Euripides, *Herakleidai* 690, σμικρὸν τὸ σὸν σήκωμα προστίθης φίλοις ("slight is the burden [weight] you add to your friends"); Polybios 8.5.91 and 21.27.4, σηκώματα μολύβδινα; cf. Pollux 4.172.

Τὸ σήκωμα can also designate a device for measuring capacity of both dry and wet commodities, as in *I. Délos* 1820, σήκωμα σιτηροῦ ἡμεδίμνου, inscribed on the side of a stone table dedicated to Apollo by an Athenian ἐπιμελητὴς Δήλου. This object, like many other examples excavated on Delos,[116] Thasos,[117] Crete,[118] and elsewhere,[119] has a circular depression sunk into its top surface into which was poured the commodity to be measured, in this case grain. Some *sekomata* also bear inscriptions on their upper surface, next to the circular depression(s), recording the volume(s) in question; e.g., χοῦς, μό[διος], [ἡμ]ίεκτον, κοτύλη on *IG* V.1.1156, from Gytheion, and others collected by Deonna (1913, page 171, note 1) and van Effenterre (1960).[120] Closer to home is a fragment of an inscribed marble measuring table among the inscriptions at Eleusis that Clinton is preparing for publication (inv. no. E659). It bears inscriptions next to the funnel-like cavities on the top, one of which shows that it was made for dry measures, probably grain.[121]

Measuring tables of this kind are known also from Athens and the Peiraieus,[122] and σηκώματα figure prominently in the well-known Athenian decree of ca. 100 B.C. regulating weights and

[115] See Nikophon's law on silver coinage, *SEG* XXVI 72, lines 20–22, with the commentary in Stroud 1974, pp. 179–181; Gauthier 1976, pp. 80–83; Velissaropoulos 1980, p. 249.

[116] *I. Délos* 1827–1829; 1847. According to Deonna (1938, p. 167), who catalogued at least fifty-eight examples, more such stone tables have been discovered on the commercial island of Delos than at any other place in the Greek and Roman world. See also Rauh 1993, pp. 12–13, 17.

[117] Pouilloux 1954, pp. 405–407, no. 153; Dunant and Pouilloux 1958, pp. 101–102, no. 194.

[118] van Effenterre 1960.

[119] The excellent article by Michon in *DictAnt* IV.2, *s.v.* σήκωμα, with copious examples, has been supplemented by Tarbell 1891; Deonna 1913; 1938; Wilhelm 1915; Guarducci 1969, pp. 469–472.

[120] In late papyri from Oxyrhynchos, the word σηκώματα designates wine jars of a fixed capacity: *POxy* XIV.1720, 4th century A.C.; XVI.1896, lines 19–20, A.D. 577.

[121] I am indebted to Clinton for showing me a copy of a draft of his edition of this inscription and for much useful discussion about *sekomata*.

[122] Koumanoudes 1862, cols. 23–24 (uninscribed); Michon in *DictAnt* IV.2, p. 1178; Tarbell 1891, p. 440, note 1 (uninscribed). Michon's suggestion, followed by Dittenberger in *IG* III 98, that the inscribed base first published by Koumanoudes (1860, p. 15, no. 26 = *IG* II^2 2886) belonged to a measuring table of this sort, is not borne out by any appropriate cuttings on the top surface. Scholars have often compared its text, [– – – Εὐ]πυρίδης ἀγορανόμο[ς γενόμενος– – –] | [– – – τὸν] ζυγὸν καὶ τὰ μέτρα ἀνέ[θηκεν] (" . . . of Eupyridai dedicated the weight and the measures after serving as *agoranomos*"), with that on the *sekoma* *IG* V.1.1156, from Gytheion. I am grateful to C. Kritzas, Director of the Epigraphical Museum, for examining *IG* II^2 2886 (EM 10573) for me and providing photographs and excellent notes on its physical condition.

measures, *IG* II^2 1013 + *SEG* XXIV 147–148.[123] The missing beginning of this document, in which its purpose was presumably stated, has added to the difficulties of its interpretation. Even though Austin has provided a helpful translation and Breglia Pulci Doria has illuminated its commercial setting,[124] many practical details in this decree remain obscure. A full commentary on Pleket's convenient composite text is badly needed.[125] On the three passages in this document (lines 7–8, 40–43, and 54–55) in which the word σηκώματα appears, Boeckh (*CIG* I 123) observed long ago *qua voce et pondera et reliquae mensurae comprehenduntur*. Here it appears to have the meaning of "exemplar" or "copy" of official standards of both weight and volume.

In trying to decide if τὸ σήκωμα in lines 25–27 of Agyrrhios' law means a weight or a measuring table, we turn to the rare verb σηκόω, which governs it in line 26. In *IG* II^2 1407, lines 40–41, this verb clearly means "to weigh": [στα]|θμία χαλκᾶ *v* ΔΙΙ *v* ἃ ὁ δῆμος σηκῶσαι ἐψ[ηφίσατο] ("twelve bronze weights which the demos decreed to weigh").[126] In an obscure passage in Plutarch, *Moralia* 928D, however, though it is surrounded by words connected with weight, the verb seems to mean "arrange," "dispose," and to be almost synonymous with κοσμέω: οὕτως οὐ ταῖς ῥοπαῖς σεσήκωται κατὰ βάρος καὶ κουφότητα τῶν σωμάτων ἕκαστον ἀλλ' ἑτέρῳ λόγῳ κεκόσμηται ("It is not in this way by the tilting [of the scales] according to heaviness or lightness that each of the [heavenly] bodies is arranged, but their disposition is due to another principle"). *Pace* H. Cherniss, Loeb edition vol. XII, p. 97.

This sense of "arrange" or "dispose" for the verb σηκόω may also be present in the First Augustan Edict from Kyrene, 7/6 B.C., lines 24–26, which defines the process of judicial allotment:[127] τότε σηκωθεισῶν τῶν σφαιρῶν καὶ ἐπιγραφέντων αὐταῖς τῶν ὀνομάτων. In commenting on this passage, Nicolet follows the translation *aequatis pilis*, printed by Riccobono, and observes, "Il est vrai que σηκόω veut dire 'peser;' mais je pense qu'ici le mot traduit *aequare*, car s'agissant de boules (*pilae*, σφαῖραι) la forme compte autant que le poids."[128] He cites other examples of *aequare* used of allotment *pilae* in Latin inscriptions.

Have we then to take τὸ σ⟨ή⟩κωμα ἐπὶ TH[. .]|ΩN<H>I σηκώσας as "weighing the weight" or perhaps as "arranging, disposing, equalizing the measuring table"? Among possible restorations of the object of the preposition ἐπί, we might propose ἐπὶ τῆ[ι χ]|ών⟨η⟩ι. A funnel (χώνη) probably was required to measure and weigh the grain, surely one made of metal like the bronze funnel from Assos in *Syll.*^{3} 945, line 7.[129] The term χώνη might also have designated the conical-shaped metal receptacle that Deonna plausibly argued was set down into the circular depressions on the upper surface of some stone measuring tables.[130] Hence, the *priamenos*, supervised by the ten men, might have determined weight or volume "by means of the χώνη," that is, by filling this conical receptacle.

Alternatively, ἐπὶ τῆ[ι ζ]|ών⟨η⟩ι, "at the rim," might be possible. If the lawgiver envisaged measuring grain in a stone σήκωμα like the numerous surviving examples mentioned above, and σηκώσας could here mean something like *aequare*, the clause could then supply essential information about how the grain is to be measured. The *priamenos* is to level off his wheat and barley ἐπὶ τῆ[ι ζ]|ών⟨η⟩ι, not heap it up above the rim or lip. Ζώνη could conceivably describe a raised ring around the top circumference of the depression in the upper surface of a stone measuring table.

[123] Day (1942, p. 112, note 350) suggested a connection between the *sekomata* in this decree and the contemporaneous dedication of several stone measuring tables on Delos.

[124] For these references see *SEG* XXXV 103.

[125] Pleket 1964, pp. 22–27, no. 14.

[126] For σηκόω meaning "lift" or "raise" (modern Greek σηκώνω), see Veyne 1985, pp. 621–624; *SEG* XXXV 967.

[127] Riccobono et al. 1941, pp. 403–407, no. 68.

[128] Nicolet 1991, p. 497.

[129] Amyx 1958, pp. 255–259, provides a useful discussion of funnels.

[130] Deonna 1913, p. 170; 1938, pp. 169–172. In publishing an uninscribed stone measuring table from Athens, Koumanoudes (1862, cols. 23–24), observed, «αἱ πέντε τρύπαι, οὖσαι χωνοειδεῖς καὶ βαθμιαίως γινόμεναι»; cf. B. Leonardos, «Λυκοσούρας Σήκωμα 'Επιγεγραμμένον», Ἀρχ 'Εφ 1899, col. 49: «'Επὶ τῆς . . . ἐπιφανείας σῴζεται λείψανον κοιλότητος μειουμένης χοανοειδῶς πως πρὸς τὰ κάτω».

Several surviving *sekomata* have such raised rims around the depressions on their upper surface to guard against spillage and overflow.[131]

Less attractive, perhaps, is the supplement ἐπὶ τῆ[ι *vacat*]|ὠν⟨ῆ⟩ι, "in accordance with the contract," "following the terms of the agreement."[132] Such a phrase would seem out of place between τὸ σ⟨ή⟩κωμα and σηκώσας, both of which probably apply to a specific physical process. We would also have to postulate an uninscribed space at the end of line 25, something not found elsewhere in the preserved portion of the main body of the text. I have not found that ἐπι-τη[. .]|ων⟨η⟩ι or ἐπὶ τη[. .]|ων⟨η⟩ι[133] or ἐπ' ἰτη[. .]|ων⟨η⟩ι yield satisfactory restorations.

The procedure of weighing and/or measuring grain briefly defined in lines 25–27 recalls a passage in the first bronze tablet from Herakleia, *IG* XIV 645, I, lines 102–104, late 4th century B.C., ἀπαξόντι ἐς τὸν δαμόσιον ῥογὸν καὶ παρμετρησόντι τοῖς σιταγέρταις τοῖς ἐπὶ τῶν ἐτέων τῶι δαμοσίωι χοὶ μεστὼς τὼς χοῦς κριθᾶς κοθαρᾶς δοκίμας ("they will convey to the public granary and they will measure out in concert with the grain commissioners holding office each year, by means of the public measure of a chous, full choes of barley clean and tested") (Uguzzoni and Ghinatti 1968, *ad loc.*).

In his forthcoming publication of the inscriptions from Eleusis, Clinton will suggest that the terms ἐπίμετρα τῶι διαμετρηθέντι σ[ί]τωι (lines 254, 281, 285) and ἐπιβολή (lines 285, 298), which accompany totals of measures of wheat and barley in *IG* II^2 1672, represent an amount of grain added to that which had been measured out and turned over, probably to compensate for the lack of compactment in the measuring process. It was the obligation of the vendor or someone making a payment in grain (such as Hypereides paying his rent in lines 252–255 of this inscription and the Attic tribes and klerouchoi dedicating their ἀπαρχαί) to provide this extra grain in order to reach the proper number of compacted *medimnoi*. In this document of 329/8 B.C., the *epimetra* vary in relation to the measured amount of grain from 9.37% (line 254) to 5.77% (line 281) to 5.36% (line 285). Although the *priamenoi* in the grain-tax law are also in a sense making a payment of grain to the *polis*, nothing is said in our text about any extra grain. Does this mean that in 374/3 B.C. the practice of adding the *epimetron* was not yet followed in weighing and measuring of grain owed to the state? Or is it possible to infer from the lawgiver's silence on the *epimetron* that this requirement already formed part of the ἐμπορικοὶ νόμοι followed by οἱ ἄλλοι ἔμποροι and that prescription was here unnecessary?[134]

Measuring and weighing grain is such a complex and potentially contentious process that we might have expected Agyrrhios' law or, more probably, the ἐμπορικοὶ νόμοι of Athens to have included specifications on such important matters as (1) whether the measures of grain were to be heaped up or leveled off, (2) was the grain shaken, stirred, or stuffed into the containers?[135] (3) from what height was the grain poured into the *sekoma* either from a sack or shoveled in from a heap on the floor and with what force? (4) was every last *medimnos* weighed and measured, or was the *priamenos* required to go through this process in front of the ten men for only a representative sample from each μερίς? (5) what was the shape (χώνη?) and height of the container into which the grain was poured? Some of these factors can affect the weight and capacity of a standard container of wheat or barley to a significant degree, perhaps as much as 25 percent.

[131] I have not found an exact example of the word ζώνη so used. Σηκόω in the sense of "to place or dispose something on or in a σήκωμα" might find a parallel in the verb τραπεζόω, "to arrange or dispose sacrificial meat and other objects on a sacrificial τράπεζα." See *IG* V.1.1390, line 86; 3447; *IG* XII.2.272; Julian, *Orations* 5.176d; Gill 1991, p. 12.

[132] For ὠνή in this sense, see Andokides 1.73, quoted above, p. 52.

[133] Lexically the most attractive supplement might be ἐπὶ τη[λι]|ῶν I I I. For τηλία see Pritchett 1956a, p. 315.

[134] There is no mention of ἐπίμετρον or ἐπιβολή in the 5th-century B.C. First-Fruits Decree, *IG* I^3 78, which prescribes the proportions of the harvest in *medimnoi* of wheat and barley each farmer owed to the goddesses as his ἀπαρχή.

[135] Moritz (1958, pp. 90, 184–186) briefly discusses the problems of compactment in measuring flour.

For a fascinating account of the complexity and the controversy involved in measuring grain in 18th-century Poland and France, see Kula 1986, pages 46–49, 62–67, 200–207, which demonstrates how ignorant we are about the same process in ancient Greece.[136] The topic of weighing and measuring grain in the Greek city-states might repay closer study. To judge from the multiplicity of standards in weight, measures, coinage, calendars, and other Greek institutions, we might expect to find as much local variation—and as much controversy—as in Kula's examples from Poland and France.

This topic requires more detailed discussion than I can give it here. I leave it with the following questions. Is it significant that the ten men supervise the weighing of the grain (ἀποστησάμενοι, middle, line 40), whereas each *priamenos* is explicitly given the right in the law to perform this process himself (ἀποστήσει, lines 21–22, 24–25; σηκώσας, line 26)? Could this syntax imply a specific safeguard for the *priamenos* against suspected attempts on the part of the ten men somehow to "tilt" the measuring devices in favor of the *polis*?[137] Could one of the reasons why most Greek measuring tables consist of thick, heavy slabs of marble or limestone have been to reduce shaking or vibrations that would cause the grain to settle in the containers?

In measuring the state's grain, the *priamenoi* and the ten men may have been assisted by the public slaves called προμετρητεῖς, who are attested at Athens before the last third of the 4th century B.C. by, e.g., Deinarchos, *Against Agasikles* F 7.2, Σκύθου τοίνυν τοῦ προμετρητοῦ ἦν υἱός· ἐν δημοσίοις γέγονε καὶ αὐτὸς ἐν τῇ ἀγορᾷ προμετρῶν διατετέλεκε καὶ ὑμεῖς ἐκλαμβανόμενοι παρ' αὐτοῦ τοὺς πυροὺς διατελεῖτε ("Now he was the son of the *prometretes* Skythes and he was himself one of the public slaves and has continued to be a *prometretes* in the Agora and you are yourselves in the custom of taking your wheat from him"), from Harpokration and *Souda Lexicon*, *s.v.* προμετρητής. See also Bekker, *Anecdota Graeca* 1.290.33–291.2, Προμετρηταί· ἄρχοντές τινες ἐνιαύσιοι ἦσαν, οἱ τῷ δικαίῳ μέτρῳ διαμετροῦντες τὰ ὄσπρια καὶ τοὺς πυροὺς ἐν τῇ ἀγορᾷ ("*Prometretai:* they were certain magistrates holding office for one year who measured out pulse and wheat in the Agora with a just measure").[138] *IG* II^2 1672, lines 291 and 299 record the μισθός of the προμετρητής who presumably helped measure out the wheat and barley sent as *aparche* for Eleusinian Demeter and Kore in 329/8 B.C.[139]

Finally, in assimilating the procedure of weighing and measuring grain in his law to that followed by οἱ ἔμποροι, Agyrrhios inserts the adjective ἄλλοι in lines 26–27, καθάπερ οἱ ἄλλοι ἔμποροι. I take ἄλλοι here in the inclusive sense to mean "just as the other emporoi," not exclusively as "the emporoi, in addition" or "just as the emporoi also do."[140] Finkelstein's classic study (1935) of ἔμπορος, ναύκληρος, and κάπηλος has warned us all against rigid classification of these terms on the basis of both activity and venue. It is clear from the context, however, that Agyrrhios conceived of οἱ ἄλλοι ἔμποροι in line 26 as men who had experience in measuring and weighing grain, presumably when they sold it. We do not know the exact origin of the grain they sold, but *emporoi* were merchants who traveled across the seas to import grain into Athens. This was also what Agyrrhios prescribed for the *priamenoi* of the δωδεκάτη τοῦ σίτου τῶν νήσων. I infer from the formulation οἱ ἄλλοι ἔμποροι in line 26 that many of the *priamenoi* envisaged by the lawgiver were

[136] For an update with valuable bibliography and a useful section on Roman grain measuring, see Corbier 1984, pp. 72–74.

[137] Kula 1986, pp. 151–152, 200–207: "The matter was also of importance when tax grain was being remeasured; hence the peasants actually demanded that they themselves, and not the lord's clerks, should do the measuring and striking" (p. 201).

[138] Cf. Rhodes 1981, p. 577. These passages should be added to the testimonia on the sale of wheat in the Athenian Agora collected by Wycherley in *Agora* III, pp. 193–194. On this topic, see note 169.

[139] I do not find persuasive the attempt of Oikonomides (1964, pp. 88–89) to restore the Athenian inscription *Hesperia* 33, 1964, pp. 225–226, no. 73 (*SEG* XXI 674) as providing evidence for the measuring of grain, although S. N. Koumanoudes raised the attractive possibility that it may show a connection between the sanctuaries of Theseus and public σῖτος in the 2nd century B.C.; see *SEG* XXVI 213.

[140] For this second sense, which is of course well attested, cf., e.g., Plato, *Gorgias* 473C7–D1, ὑπὸ τῶν πολιτῶν καὶ τῶν ἄλλων ξένων, with E. R. Dodd's good note on 447C3.

also grain merchants. For similar wording cf. [Demosthenes] 52.3, Λύκων (an ἔμπορος) . . . τῇ τραπέζῃ τῇ τοῦ πατρὸς ἐχρῆτο, ὥσπερ καὶ οἱ ἄλλοι ἔμποροι ("Lykon, a merchant, employed the bank of his father just as the other merchants did').[141] For additional discussion on the experience of the *priamenoi* in the grain trade, see pages 111–115.

LINES 27–29

προκαταβολὴν οὐ θήσει ὁ πριάμε[ν]ο[ς ἀ]-
λλ' ἐπώνια καὶ κηρύκεια κατὰ τὴν [μ]ερ[ίδ]-
α εἴκοσι δραχμ⟨ά⟩ς

The buyer of the tax will not make a down payment but (he will pay) sales taxes and herald's fees of twenty drachmai per portion

Agyrrhios excuses the *priamenos* from the down payment (*prokatabole*) that was included in most contracts between the *polis* and a tax-farmer.[142] Down payments were also sometimes required of those purchasing property sold at public auction; see below, pages 62–64. Exemption from a prepayment for the collectors of the *dodekate* on the grain from the islands might be explained by the conjecture that they are contracting to render payment of the tax to the *polis* not in cash but in kind. They are not required to provide a certain number of *medimnoi* of wheat and barley "up front" in advance to qualify for the right to collect the *dodekate* in the islands. Unlike most other tax-farmers and lessors of public property, they are also presumably exempt from installment payments.[143] For the *priamenoi* of the $8\frac{1}{3}$% tax on the grain in the islands the lawgiver specified only a single deadline, πρὸ τοῦ Μαιμακτηριῶνος μηνός, line 48, by which time they must have delivered all their *merides* of wheat and barley to the Aiakeion. For the προκαταβολή in lines 55–56, see below, pages 78–81.

Although excused from a *prokatabole*, the *priamenos* is held responsible, nevertheless, for payment of ἐπώνια καὶ κηρύκεια. The former appears in other contexts as sales taxes accompanying the sale of property and commodities.[144] On the basis of the numerous examples of ἐπώνια as sales taxes on the confiscated property of the Hermokopidai in 414/3 B.C., *IG* I³ 421–430, Pritchett (1953, pages 226–230) urged that they were in this case computed not as a percentage

[141] For *emporoi*, see the helpful discussions in Isager and Hansen 1975, pp. 64–74; W. E. Thompson 1982; M. V. Hansen 1984; Pritchett 1991, p. 425, note 625; Cohen 1992, pp. 17, 120, 123, 134, following C. Mossé in translating "all the *emporoi*" in [Demosthenes] 52.3. For *emporoi* as grain-merchants, see Velissaropoulos 1980, pp. 35–37, and at Rome, Rickman 1980, pp. 142–143.

[142] *Souda Lex.* and Photios, *s.v.* προκαταβολὴ καὶ προσκαταβόλημα· τῶν τελῶν πιπρασκομένων, δύο προθέσεις ἐδίδοντο τοῖς ὠνουμένοις, ἐν αἷς ἐχρῆν εἰσενεχθῆναι τὸ ἀργύριον. ὅπερ οὖν μέρος χρημάτων πρὶν ἄρξασθαι τοῦ ἔργου εἰσφέρουσι εἰς τὸ δημόσιον, τοῦτο προκαταβολὴ καλεῖται, τὸ δὲ τῇ δευτέρᾳ προθεσμίᾳ διδόμενον προσκατάβλημα ("Down payment and supplementary payment: when the taxes were sold, they used to grant two installments to the purchasers through which the money had to be paid in. Accordingly, that portion of the money they pay into the public treasury before embarking on the project is called the down payment. That which is paid by means of the second installment is called the supplementary payment"). Cf. Bekker, *Anecdota Graeca* 1.193.7; *SEG* XII 100, line 36 (below, p. 64), down payment on a sale of property; Hallof 1990, p. 411.

[143] For installment payments for tax-collectors, see Aristotle, *Ath* 47.3, with Rhodes' 1981 good note *ad loc.* For the wide variety in the numbers and timing of payments for leases and sales of property, see, e.g., *IG* I³ 84, line 15; *SEG* XXVI 72, lines 6–7, with Stroud 1974, pp. 167–168; Boeckh 1886, I, pp. 409–414; Osborne 1988, pp. 283–284; Walbank, *Agora* XIX, pp. 162–163; Hallof 1990, pp. 407–409.

[144] E.g., *IG* II² 1579 + *SEG* XXXII 161 (*Agora* XIX, P2), lines a, b, c [10], [18], [25], d [5], 12, 18, e [3], 6, f 3, g 12; *SEG* XII 100, line 37 (below, p. 64); *SEG* XIX 132 (*Agora* XIX, P3), line [4]; *Agora* XIX, LA2, line [2]; *Hesperia* 16, 1947, pp. 155–157, no. 51 (*Agora* XIX, P53), lines 6 and 46. Still helpful on *eponion* is Boeckh 1886, I, p. 395, with Fraenkel's notes *ad loc.* Boerner's incomplete article (1907), *s.v.* 'Επώνιον was replaced by W. Schwahn (1934), *s.v.* Tele. See also Andreades 1933, pp. 144–150 (rambling and unsatisfactory); Hallof 1990, pp. 407–410, 413, with helpful bibliography: "Das ἐπώνιον ist eine in Athen auf den Verkauf vor den Poleten beschränkte, abgestufte Kaufsteuer, die der Käufer zu entrichten hat" (p. 410).

of the selling price, but on a sliding scale.[145] Other rates attested in the lexicographers are 1%, Bekker, *Anecdota Graeca* 1.255.1–4[146] and 20%, Harpokration *s.v.* (= Isaios F 43).[147] On the basis of the ἐπώνια recorded with sales on the *poletai* inscription *IG* II^2 1579 + *Hesperia* 5, 1936, pages 390–393, no. 9 (*SEG* XXXII 161; *Agora* XIX, P2), Pritchett (1956b) argued that "in the fourth century B.C." the *poletai* were using a sliding scale for computing sales taxes at a rate exactly double that used on the Attic Stelai.[148] Although his readings of this inscription, plus new fragments, confirmed Pritchett's computation of the ἐπώνια totals, Walbank (1982a) showed that the sales here recorded were those of the confiscated property of the Thirty. The inscription is dated in 402/1 B.C.; the doubling of the sales-tax rate followed the Attic Stelai by little more than a decade. How long it remained at this level cannot be determined on the basis of epigraphical evidence from the 4th century B.C., for ἐπώνια in documents of this period are now almost always combined with κηρύκεια to yield a single total, the proportions of each component being unstated. It is also likely that the modern search for a standard rate of sales tax may be misplaced. Tax rates may have fluctuated in ancient Athens as frequently as they do elsewhere today. Taxation can be a volatile agent in the economy—ancient and modern—sensitive to changing pressures both local and temporal. Hence the level of sales tax may even have been determined on an *ad hoc* basis at the time when the *polis* drew up the contract for the farming of a tax, the sale of confiscated properties, or the leasing of public lands and mining rights.

Apart from Pollux 7.15, τὰ δὲ καταβαλλόμενα ὑπὲρ τῶν πιπρασκομένων τέλη ἐπώνια λέγουσιν ("they call the money paid down on behalf of those purchasing taxes *eponia*"), the grain-tax law provides the only evidence known to me for the payment of *eponia* by tax-farmers at Athens. In our law the *priamenoi* of the *dodekate* are all required to pay the same amount, which suggests that in this instance the *eponion* was set at a flat rate, not a percentage of the sale price and possibly not even computed on a sliding scale. Together with the *kerykeion*, the *eponion* in each instance will be twenty drachmai per *meris*.

An unspecified portion of the twenty drachmai per *meris* is reserved for *kerykeia*. Sales of confiscated property, some leases, and the farming of taxes seem normally to have been conducted at Athens through public auction.[149] The *polis* charged the "purchaser" (πριάμενος) a fee (κηρύκειον) for the auctioneer (κῆρυξ).[150] We know very little about the procedure of such auctions, most of

[145] See also Pritchett 1954. For earlier computations along the same lines, see Fränkel's note 536 in Boeckh 1886, II, pp. 76–77. Hallof (1990, pp. 408–409) argues that with the exception of only two amounts, all totals of *eponia* on the Attic Stelai can be explained as computed at 1%.

[146] Hallof (1990, pp. 407–408) holds that the *eponion* is distinct from the ἑκατοστή in Aristophanes, *Wasps* 658; Theophrastos in Stobaios, *Flor.* 4.2.20, and the Attic *rationes centesimarum* documents, on which see above, note 41.

[147] *S.v.* ἐπώνια· τέλος ἐστὶ τὸ ἐπὶ τῇ ὠνῇ διδόμενον, εἴη δ' ἂν ἴσως ἡ πέμπτη· Ἰσαῖος ἐν τῷ Κατ' Ἐλπαγόρου καὶ Δημοφάνους (F 43) κἀν τῇ Πρὸς Τληπόλεμον ἀντωμοσίᾳ (F 128) ("*eponia:* it is a tax which is paid on a purchase; perhaps it might be a fifth; Isaios in his speech *Against Elpagoras and Demophanes . . . and in the affidavit in Reply to Tlepolemos*"). Boeckh (1886, I, p. 395) rejected this figure as far too high and has been vindicated by subsequent discoveries; above, note 145. For an example of *eponion* calculated on a sliding scale, not as a percentage of the sale price, see *Syll.*^{3} 1014 (Erythrai) with Dittenberger's note 3 on p. 151.

[148] For Hallof (1990, pp. 409–410) this rate is a full 2%.

[149] J. Oehler collected a number of examples of this function of κήρυκες at Athens and elsewhere (1921), *s.v.* Keryx. The Athenian *poletai* may have had their own *keryx* called πρατίας: see Pollux 7.8; Hesychios, Photios, *s.v.* ὁ τὰ δημόσια πωλῶν, κῆρυξ δήμοσιος. Hallof 1990, p. 412.

[150] Harpokration, *s.v.* κηρύκεια· ὁ διδόμενος μισθὸς τοῖς κήρυξιν ἐπὶ ταῖς γινομέναις πράσεσιν· Ἰσαῖος ἐν τῷ Κατ' Ἐλπαγόρου καὶ Δημοφάνους ("*kerykeia:* the pay granted to heralds when sales took place; Isaios in his speech *Against Elpagoras and Demophanes*"); cf. *Souda Lex. s.v.;* Bekker, *Anecdota Graeca* 1.255.2–4, κηρύκεια δὲ τὰ τῷ κήρυκι διδόμενα ὑπὲρ τοῦ κηρύττειν τὰ τέλη πιπρασκόμενα ("*kerykeia:* money granted to the herald for serving as herald in the selling of taxes"), Pollux 4.93, κηρυκεία [*sic*] τέλους τινὸς ὄνομα ("*kerykeia:* the name of a tax"). Cf. *SEG* XII 100, line 37 (below, p. 64); XIX 132 (*Agora* XIX, P3), line 4; *Hesperia* 16, 1947, pp. 149–150, no. 39 (*Agora* XIX, P45), line 3; *Hesperia* 16, 1947, pp. 155–157, no. 51 (*Agora* XIX, P53), lines [6] and 47;

which were conducted under the supervision of the *poletai*, and less about how the auctioneer's fee was normally computed. In the grain-tax law, however, the *kerykeion*, like the *eponion*, was set at a flat rate in advance.

As Hallof (1990, pages 410, 423) points out, surviving sources give no indication into what fund the *eponion* was allocated when a buyer paid it. Nor does the grain-tax law. Did Agyrrhios intend, for instance, that the *priamenoi* would pay the fee of twenty drachmai per *meris* directly to the *poletai* who would then divide it into the herald's *misthos* and the *eponion*? Or did the *poletai* turn the full sum over to, say, the *apodektai*, who would allot the herald his fee? Did the *priamenoi* themselves have to pay the twenty drachmai directly to the *apodektai*? *Souda Lexicon, s.v.* πωλητής states only that the *poletai* turned the proceeds of sales over to τὸ δημόσιον.

I have not found another example of *eponia* and *kerykeia* assessed as a specific sum in drachmai in advance of the "sale" of a tax contract at Athens. This perhaps suggests that Agyrrhios may not have had a normal auction in mind for the sale of the right to collect the *dodekate* and the *pentekoste* on the grain in the islands. As we have seen above (pages 40–41), he also clearly anticipated awarding contracts to several successful bidders and not just to a single individual or to one *symmoria* of six *priamenoi* who might have made the highest bid. Nor was the normal requirement of προκαταβολή in effect for successful bidders. Can we infer from the uniform charge of twenty drachmai per *meris* that Agyrrhios regarded ἐπώνια καὶ κηρύκεια for the *priamenoi* in his law more as a flat fee required in return for the authorization by the *polis* of the right to collect the *dodekate* rather than as a sum to be computed on the basis of an (unknown) winning bid?

Although it is clear from the presence of ἐπώνια καὶ κηρύκεια κατὰ τὴν [μ]ερ[ίδ]α εἴκοσι δραχμ⟨ά⟩ς in lines 28–29 that a uniform amount was set by this law and not on an *ad hoc* basis by the *boule* working with the *poletai*, the origin of the figure remains obscure. How did the lawgiver arrive at it? In the sales of confiscated property of the Thirty in 402/1 B.C., we have seen that the rate of sales tax was roughly 2%. Agyrrhios' fee of twenty drachmai is 2% of 1,000 drachmai, but I do not see the significance of the latter sum in the context of the grain-tax law. It is difficult to imagine how twenty drachmai can be a percentage or part of a sliding-scale figure tied to the eventual sale of the people's grain in the Agora by the ten men. As lines 44–46 tell us explicitly, the price at which the wheat and barley from the islands are to be sold will not be set by the *ekklesia* until some future unstated time. The price cannot be known at the time when the *dodekate* and *pentekoste* are "sold." If 1,000 drachmai is an estimate of what public sale of 500 *medimnoi* of the people's grain will fetch, then the lawgiver is probably anticipating very low prices for wheat and barley. So perhaps we should abandon the example of 2% based on the sales of the confiscated property of the Thirty.

John Camp has suggested to me the possibility of using as *exempli gratia* figures the prices of six drachmai per *medimnos* of wheat and three drachmai per *medimnos* of barley attested in *IG* II^2 1672 (see above, pages 32–33) to make the following computation:

1 *meris*	100 *medimnoi* of wheat @ 6 dr. per *medimnos*	600 dr.
	400 *medimnoi* of barley @ 3 dr. per *medimnos*	1,200 dr.
	Total	1,800 dr.

This would yield for twenty drachmai per *meris* (*eponia* and *kerykeia*) a percentage fairly close to 1%, which is a rate attested in the Attic Stelai and in Bekker, *Anecdota Graeca* 1.255.2; see also the argument of Hallof (1990, pages 408–409).

It is probably safer at this point to conclude that we do not know the lawgiver's motive for setting the fee for *eponia* and *kerykeia* at twenty drachmai per *meris*.

Langdon, *Agora* XIX, p. 58; Hallof 1990, pp. 411–412;
Langdon 1994, pp. 262–263.

All three of the key words in lines 27–28 of our law—προκαταβολή, ἐπώνια, and κηρύκεια—appear together in a *poletai* document of 367/6 B.C., *SEG* XII 100, lines 36–37 (*Agora* XIX, P5), τούτο τὴν προκαταβολὴν τὸ πέμπτον μέρος ἔχει ἡ πόλις καὶ τὰ ἐπώνια καὶ τὰ κηρύκεια ("of this [sum] the *polis* has as a down payment the fifth part plus *eponia* and *kerykeia*"). On the basis of this passage and two other sales of property in the *poletai* account *Agora* XIX, P26, Hallof (1990) has argued that in sales of confiscated property (and to some extent in farming out mining contracts), the *poletai* did not conduct a free and open auction aimed at producing the highest possible bid to bring in the most revenue for the *polis*. Buyers were preselected; prices were often fixed in advance for a variety of reasons. At first glance Agyrrhios' grain-tax law with its fixed fee of twenty drachmai for *eponia* and *kerykeia* might appear to lend support to this theory that the *poletai* merely presided over a rigged "auction."

Langdon (1994), however, mounts a spirited defense of open bidding in all the activities of the *poletai*—selling confiscated property, leasing mining contracts, and farming taxes. For our purposes here I restrict myself to the last category, for which, in my view, Langdon presents a compelling case.

Although the auction and the selling of the *dodekate* on the grain of the islands probably had, as we have seen, some unusual features, I see no reason to doubt that the *boule* and the *poletai* established first how many *merides* of grain would be presented for sale. This number could even have been made known when the date of the sale was publicly announced in Athens. On the appointed day, in the presence of the *boule*, and assisted by their *keryx*, the *poletai* probably auctioned off the right to collect the grain-tax. Grain merchants—either singly or in *symmoriai* of six men each—bid for one or more *meris*. For each *meris* of wheat and barley that they contracted to deliver to the Aiakeion before the beginning of Maimakterion, they paid a fee of twenty drachmai and produced two guarantors approved by the *boule*. The awarding of the contract by the *polis* to each *priamenos* or *symmoria*, together with the number of *merides* involved in each transaction and the names of the guarantors, was probably publicly announced by the herald. As I suggested above (pages 47–48), the *poletai* turned over a copy of this essential information to the *boule* and also to the ten men who had to supervise the timely delivery of the grain to the Aiakeion, where it was weighed, measured, and tested for quality.

LINES 29–31

ἐγγυητ⟨ὰ⟩ς καταστήσ[ε]-
ι ὁ πριάμενος δύο κατὰ τὴμ μερίδα ἀξι[ό]-
χρεως, οὓς ἂν ἡ βουλὴ δοκιμάσηι

The buyer of the tax will provide per portion two solvent guarantors, whomever the Council has approved

For guarantors of tax contracts, see Andokides 1.133–134 and Plutarch, *Alkibiades* 5, both quoted and discussed above, pages 19, 30, and Xenophon, *Poroi* 4.20, ὅπως γε μὴν τὰ ὠνηθέντα σῴζηται, τῷ δημοσίῳ ἔστι λαμβάνειν ἐγγύους παρὰ τῶν μισθουμένων, ὥσπερ καὶ παρὰ τῶν ὠνουμένων τὰ τέλη ("To make very sure that the [slaves] who have been purchased are protected, it is possible for the public treasury to take guarantors from the lessees just as it also does from those who purchase [the right to collect] taxes"), with Gauthier's good note (1976, pages 148–149). For the wording, see Aristophanes, *Ekklesiazousai* 1064–1065, ἐγγυητάς σοι καταστήσω δύο| ἀξιόχρεως ("I will provide you with two solvent guarantors"); Plato, *Laws* 9.871E3–5, ὁ δὲ παρεχέτω τοὺς ἐγγυητὰς ἀξιόχρεως οὓς ἂν ἡ τῶν περὶ ταῦτα δικαστῶν ἀρχὴ κρίνῃ ("let him provide solvent guarantors whomever the magistracy of jurors dealing with these matters shall judge fit"); 11.914D3; *Apology* 38B9; Demosthenes 24.144. See MacDowell 1962, page 63.

Walbank (*Agora* XIX, page 163, with earlier references) has plausibly suggested that guarantors for public leases were always Athenian citizens. This is a very likely inference also for the guarantors in our lines 29–31 who underwent scrutiny before the *boule*. If non-Athenian *priamenoi* did not produce the grain in accordance with the provisions of the contract, the *polis* probably had to attach the property of the guarantors. The adjective ἀξιόχρεως here has its full etymological force, "worthy of the debt," that is, possessing resources capable of covering the liability a guarantor undertakes, "solvent"; see Gofas 1993, page 185, note 116. For the action, cf. *Agora* XIX, P26 (*IG* II^2 1582+), lines 462–495 and the Samian Grain Law, *Syll.*^{3} 976, lines 68–69, ἐὰν δέ τι ἐνλίπῃ, τὴν πρᾶξιν| ποιησάσθω ἐκ τοῦ ἐγγύου ("if there is any deficiency, let it be exacted from the guarantor").

For speculation about how the number of guarantors per *meris* might have affected the form of the auction of the tax, see below, commentary on lines 31–36 and pages 110–111.

LINES 31–36

συμ[μορ]-
ία ἔσται ἡ μερὶς τρισχίλιοι μέδιμ̣[νοι],
ἓξ ἄνδρες· ἡ πόλις πράξει τὴν συμμορ[ία]-
ν τὸν σῖτον κ⟨α⟩ὶ παρ' ἑνὸς καὶ παρ' ἁπάν[τω]-
ν τῶν ἐν τῆι συμμορίαι ὄντων, ἕως ἃ̣ν τ[ὰ α]-
ὑτῆς ἀπολάβηι

The portion held by a group of six men (symmoria) *will consist of 3,000 measures* (medimnoi). *The* polis *will exact the grain from the group* (symmoria) *both from one man and from all who are in the group, until it recovers what belongs to it*

Agyrrhios, who earlier in his career had himself led (ἀρχώνης) a company of joint tax-collectors (above, pages 19–20), here spells out the terms by which six men may together contract with the *polis* to collect an expanded *meris* of 3,000 *medimnoi*, six times as large as the single *meris* of lines 8–10. Presumably—although the law does not say so explicitly—the same proportions of wheat to barley applied as in the smaller, or single, *meris* of 400 *medimnoi* of barley and 100 of wheat. It is possible also that the size of the συμμορία and its expanded *meris* reflect the carrying capacity of a normal grain ship of the Classical period; see Velissaropoulos 1980, pages 57–67; Gofas 1993, pages 212–214. Thus Agyrrhios appears to provide for joint speculation by six *priamenoi* possibly on a smaller scale than some of the other *emporoi*, with shared risks and expenses. He may have wanted to attract a wider range of bidders who could enter into the speculation with a minimum of initial investment.

Although multiple partnerships for tax-farmers are well attested at Athens, Agyrrhios' *symmoria* of six men may have been an innovation, if we are right in concluding that he expected many of the *priamenoi* of the *dodekate* to be *emporoi*. Apparently, *emporoi* seldom worked in groups of this size at Athens. According to Isager and Hansen's findings (1975, page 74), "not only lenders but *emporoi* as well often united in pairs [eight examples cited from the speeches of Demosthenes] but larger partnerships are unknown." A tax-farming company of six men for the *dodekate* on the grain from the islands would have facilitated specialized division of labor among its members. Not all six, surely, would have had to sail to the islands. Some members, for instance, may have supervised transport of the grain from the Peiraieus to the city or taken charge of other tasks. Is our law providing a glimpse here of fairly sophisticated business practices, carried out by entrepreneurial, private companies in the early 4th century B.C.? To my knowledge we do not have evidence in Greece this early for corporations of merchants as highly organized as the Roman *collegia*, on which see, e.g., Rickman 1980, pages 87–92, 226–230.

The law does not state whether the *eponia* and *kerykeia* for the συμμορία μερίς are to be assessed at twenty or 120 drachmai nor whether two or twelve guarantors are required.

Although no other surviving source earlier than 374 B.C. appears to apply the word συμμορία to a company or association of partners in a commercial or financial context, companies of tax-farmers at Athens are attested well before the date of this law, as we have seen above, pages 19–20. It is even possible that the two associations headed by Agyrrhios and Andokides in 402/1 B.C., which farmed the *pentekoste*, might have been called *symmoriai*, although our main source, Andokides

1.133–134, does not use this word, or indeed any technical term to describe them (συνέστησαν πάντες and οἱ μετασχόντες).

If we can trust the testimony of the Atthidographer Philochoros, writing before the 260s B.C., the Athenians were first divided into συμμορίαι, for an unstated purpose, in the archonship of Nausinikos, 378/7 B.C. (*FGrH* 328 F 41). Most scholars have linked their introduction to a reform of the system of collecting the *eisphora* tax, suggested by Demosthenes 27.7, and to the foundation of the Second Athenian Confederacy, *IG* II^2 43.[151] Thus, Agyrrhios' use of the term συμμορία for a company of six collectors of the *dodekate* tax in his law would have had a recent antecedent. In a frustratingly imprecise discussion of the Archaic Athenian ναυκραρίαι, Kleidemos (*FGrH* 323 F 8) compares them to the συμμορίαι of his day, which he says numbered 100. Conjectures as to the date and functions of these *symmoriai* abound.[152]

At Athens, of course, *symmoriai* are most familiar as the groups of wealthy contributors who formed the intricate system through which the law of Periandros, 358/7 B.C., and subsequent legislation organized the financial support of the fleet. These later naval *symmoriai*, the topic of Demosthenes 14 and other speeches, will not concern us here, despite the similarity in nomenclature.[153] Nor, happily, need we enter into the vigorous debate, at least as old as Boeckh 1886, I, pages 599–623, as to whether there were two different kinds of *symmoriai: eisphora-symmoriai* and trierarchic-*symmoriai*, with different numbers, organization, and personnel, or, as Ruschenbusch (1978, page 277) maintains, "diese ganze Theorie von den zwei Arten von Symmorien ist falsch," for "es in Athen nur ein einziges Symmoriensystem gegeben hat" (page 284).[154] The new evidence for the *symmoria* of six men in the grain-tax law, however, demonstrates that Ruschenbusch's topic sentence, "Symmorien sind Abteilungen für die Zahlung der Eisphora und die Leistung der Trierarchie" (page 275), is an oversimplification. We see now that the term συμμορία in 4th-century B.C. Athens cannot be so tightly circumscribed.

While providing for and encouraging the formation of tax-collecting cartels, the law is careful to restrict them in size and to some extent in their freedom of operation. It would be wrong, I think, to regard this provision of the law as a departure from the panoply of Athenian regulations on the grain trade which aimed at keeping hoarding and speculation on prices at a minimum. Accordingly, the *polis* will exact the grain from the members of the *symmoria* both individually and from all of them as a whole, ἡ πόλις πράξει τὴν συμμορ[ία]ν τὸν σῖτον κ⟨α⟩ὶ παρ' ἑνὸς καὶ παρ' ἁπάν[τω]ν τῶν ἐν τῆι συμμορίαι ὄντων, lines 33–35. For this use of πράττειν in a tax context, with two accusatives, see, e.g., *PCG* II, Alexis F 265 (= Athenaios 1.21d), οὗ μήτε πράττεται τέλος| μηδεὶς γὰρ ἡμᾶς ("for which no one is exacting a tax from us"). Likewise, in the context of a contract, see *IG* VII 3073, lines 3–4, ἅπαν|τα (ἀργύριον) πράξ[ουσιν] οἱ ναοποιοὶ τὸν ἐργώνην καὶ τοὺς ἐγγύους ("the temple-builders will exact all the money from the contractor and his guarantors"). See also Migeotte 1984, nos. 12, line 49 ("on trouve indifféremment l'actif ou le moyen," copious examples); 13, line 45.

In this final clause of what I have called the contract part of the law, lines 8–36, the verb (πράξει) continues to be expressed in the future indicative. It has as its subject, for the second time (cf. lines 20–21, ἐνοίκιον οὐ πράξει [ἡ π]|όλις τοὺς πριαμένους), one of the two parties to the contract, ἡ πόλις. Just as the law did not earlier designate any specific officials who were to act on behalf of the *polis* in providing the Aiakeion as a repository for the grain (lines 15–16), so here it is merely ἡ πόλις that will exact the *sitos* from the *symmoria*. Throughout the terms of the contract

[151] From the enormous bibliography on this topic, see, e.g., Thomsen 1964, pp. 45–104, 194–206, 226–228, who does not, however, accept Philochoros' statement; Ruschenbusch 1978; Brun 1983, pp. 3–73; Rhodes 1982; MacDowell 1986; Ruschenbusch 1987; Gabrielsen 1990.

[152] E.g., Jacoby's discussion on Kleidemos, *FGrH* 323 F 8, and those cited in the previous note.

[153] For excellent succinct summaries see Rhodes 1981, pp. 679–682; MacDowell 1990, pp. 367–369, 371–373, 375–376; and now at full length Gabrielsen 1994.

[154] For this same view see Mossé 1979; MacDowell 1986; Ruschenbusch 1990.

section, the law has consistently designated the other party as ὁ πριάμενος (lines 11–12, 17–18, 22, 27, 30; plural in line 21), but here in lines 32–36 he is replaced by the six men of a *symmoria*. They too, however, are also *priamenoi*. It is significant, I think, that at this point Agyrrhios inserts an exactment clause: the *polis* will exact its grain from the *symmoria*, both from each individual member and from all of them who belong to it. The *polis* will carry out the exactment until it secures what belongs to it, ἕως ἂν τ[ὰ α]ὑτῆς ἀπολάβηι (lines 35–36). For the wording, cf. Damonides' advice to Perikles on jury pay, διδόναι τοῖς πολλοῖς τὰ αὑτῶν ("to give the majority their own"), Aristotle, *Ath* 27.4. In this instance τ[ὰ α]ὑτῆς will consist of 3,000 *medimnoi* for each *meris* of a *symmoria*. For ἀπολαμβάνω in the sense of "recover what is one's own," see *IG* II^2 212, lines 54–58, ὅπ[ως| ἂ]ν ἀπολάβωσι . . . ὅ[πως ἂ|ν] ἀπολα[β]όντες τὰ χρήματα. For the reflexive τ[ὰ α]ὑτῆς, see Henry 1996, pages 117–119.

Earlier in the law an exactment clause of this type would have been superfluous, since the *polis* has so far been consistently interacting with one *priamenos* at a time, who is solely responsible, together with his guarantors, for producing the *merides* of grain for which he is awarded the contract. With a *symmoria* of six men, however, the target of the *polis*' exactment becomes more diffuse. In a discussion of "the absence of any concept akin to the modern legal notion of the partnership or corporation," Harris (1989) has aptly cited an excellent later parallel for the wording of the exactment clause in our law. Discussing a συγγραφή entered into by the lenders Dareios and Pamphilos on the one hand and the borrowers Dionysodoros and Parmeniskos on the other, the speaker in [Demosthenes] 56.45 observes that all four agreed to a penalty clause in the event of default on the repayment of the loan. Explicitly written into the contract was the additional provision that the lenders could exact this penalty from either or both of the borrowers, τὴν δὲ πρᾶξιν εἶναι καὶ ἐξ ἑνὸς καὶ ἐξ ἀμφοῖν ("the exactment is to be both from one and from the two of them"). Harris has urged that it was necessary for the lenders to appeal to this clause in the contract in order to proceed against Dionysodoros, as they did, when it was Parmeniskos who was apparently in default, because "Athenian law did not consider that joint borrowers formed a single partnership and were thus jointly liable for any breach of the agreement."[155]

So too in the grain-tax law, the explicit insertion of the exactment clause immediately following the terms under which a *symmoria* of six men may participate in the farming of the *dodekate* probably indicates the need to specify that the *polis* can and will proceed to exact its grain from all six members either individually or as a group. That is, even though this is not strictly a penalty clause, it may still demonstrate the principle that Athenian law did not *automatically* recognize joint liability in the case of joint or multiple parties to a contract. It was necessary to write such liability explicitly into the contract. This nuance is important. Enterprising lenders and merchants clearly found a way to protect their interests privately, and it is striking to see the *polis*, in the person of the veteran businessman who proposed this law, explicitly stating how it will protect its own interests.

For the later, and more complex, exactment clause in contracts preserved on papyri, see Wolff 1941; see also Simon 1964.

[155] I am indebted to Harris for pointing this out to me and for sending me a copy of his paper. He cites a similar clause in a joint-contract for a maritime loan in [Demosthenes] 35.12, which is very detailed, παρὰ Ἀρτέμωνος καὶ Ἀπολλοδώρου ἔστω ἡ πρᾶξις τοῖς δανείσασι καὶ ἐκ τῶν τούτων ἁπάντων, καὶ ἐγγείων καὶ ναυτικῶν, πανταχοῦ ὅπου ἂν ὦσι, καθάπερ δίκην ὠφληκότων καὶ ὑπερημέρων ὄντων, καὶ ἑνὶ ἑκατέρῳ τῶν δανεισάντων καὶ ἀμφοτέροις ("let the exactment be conducted by the lenders from Artemon and Apollodoros and from all their property both on land and on sea wherever they may be just as if they had been assessed a penalty and defaulted on it, and the exactment is to be carried out by each of the two lenders individually or by both together"). Cf. also *IG* VII 3172, from Orchomenos, Migeotte 1984, no. 13, lines 106–111, ἡ δὲ πρᾶξις ἔστω ἔκ τε | αὐτῶν τῶν δανεισαμένων | καὶ ἐκ τῶν ἐγγύων καὶ ἐξ ἑνὸ[ς] | καὶ ἐκ πλειόνων καὶ ἐκ πάν|των καὶ ἐκ τῶν ὑπαρχόντων| αὐτοῖς ("let the exactment be from the borrowers themselves and from their guarantors both from one individually and from more than one and from all and from their goods").

LINES 36–39

αἱρείσθω δὲ ὁ δῆμος δ[έκ]-
⟨α⟩ ⟨ἄ⟩νδρας ἐξ Ἀθηναίων ἁπάντων ἐν τῆι [ἐκ]-
κλησίαι, ὅτανπερ τοὺς στρατηγοὺς α[ἱρ]|ῶνται

Let the Assembly elect ten men from all the Athenians in the meeting at which they elect the generals

In 1991 Hansen observed, "Every time a sizable new inscription emerges from the soil of Attica there is a good chance that the list of known boards of magistrates will acquire an addition" (Hansen 1991, page 240, which goes back to Hansen 1979b, page 48).

The grain-tax law not only brings a welcome addition to the list of known boards of Athenian magistrates, but also marks the inception of a group of ten men, οἵτινες ἐπιμελήσονται τοῦ σί[τ]ο (lines 39–40). They form a frustrating addition to our knowledge of Athenian political institutions, however, for, although this part of the law is rich in details about their duties and method of appointment, I have found no later trace of these magistrates in our surviving sources. Since they do not carry a title but are merely called δέκα ἄνδρες (lines 36–37) and οἱ αἱρεθέντες ὑπὸ τοῦ δήμου (48–49), their chances of having left some tracks in the ancient scholia, lexica, onomastica, and other antiquarian works are slim. In such sources the designation οἱ δέκα at Athens conjured up a different image after the events of 403–401 B.C., Aristotle, *Ath* 35.1; 38.1, 3; 39.6; Rhodes 1981, pages 456–461.

Indicative of the importance of these new magistrates established by Agyrrhios' law is the contemporary existence at Athens of numerous other officials who supervised the importation and sale of grain and grain products in the private sector. These included the *sitophylakes*,[156] the *agoranomoi*,[157] the *epimeletai tou emporiou*,[158] and perhaps others.[159] Our text ignores all of these. Instead, it entrusts the people's *sitos* from the islands to a completely new board of magistrates for whom this appears to be the only responsibility. The creation of this new board probably indicates that if the *dodekate* and the *pentekoste* were not themselves new taxes at this time, their collection was certainly now undergoing a radical overhaul.

That there are to be ten men on the new board is, of course, not surprising in this period of the ten tribes, but the lawgiver quickly specifies that election (αἱρείσθω, line 36), not sortition, is to be their method of appointment. Tribal representation appears also to have been no concern of the *demos*, since election is to be ἐξ Ἀθηναίων ἁπάντων (line 37).[160] I infer from the timing of the election, ἐν τῆι [ἐκ]κλησίαι ὅτανπερ τοὺς στρατηγοὺς α[ἱρ]ῶνται (lines 37–39), that the term of office was of one year's duration. It also seems to me likely, in view of the specialized duties of the ten men, the fact that they are elected, not chosen by lot, and that they are in a sense linked to the annual election of the generals, that the lawgiver may have envisaged the possibility of a man serving for two or more consecutive terms.

An indication of how seriously the Athenians regarded the election of the ten men is the decision to hold it at the same meeting of the assembly when they elected the *strategoi*. We shall examine shortly another link between the grain supply and the military in lines 54–55, below, pages 77–78. Later in the 4th century B.C. in Aristotle's day, and possibly also in 374/3 B.C.,

[156] For testimonia and discussion, see Stroud 1974, pp. 180–181; Rhodes 1981, pp. 577–579; Gauthier 1981, pp. 17–19; Migeotte, forthcoming, "Aspects." Their concern was primarily with ὁ ἐν ἀγορᾷ σῖτος, Aristotle, *Ath* 51.3.

[157] Rhodes 1981, pp. 575–576.

[158] Stroud 1974, pp. 180–181; Gauthier 1976, pp. 80–83; Velissaropoulos 1980, pp. 33–34; Rhodes 1981, p. 579; Figueira 1986, pp. 150–152. On the respective functions of the *sitophylakes* and the *epimeletai tou emporiou*, see esp. Gauthier 1981, pp. 19–28.

[159] Possibly also the *metronomoi;* see Rhodes 1981, p. 577; or even the *syllogeis tou demou*, Stroud 1974, pp. 178–179; Rhodes 1981, p. 520. For the later *sitonai* at Athens and elsewhere, see, e.g., Fantasia 1987; Couilloud-Le Dinahet 1988; Strubbe 1989; Migeotte 1991; 1995, pp. 28–29; forthcoming, "Ventes."

[160] For testimonia and discussion of the election process, see Hansen 1991, pp. 230–237.

elections of the generals were conducted by the (first) *prytaneis* after the sixth prytany in whose term there were good omens, that is, in the seventh prytany at the earliest.[161] Meritt (1970) persuasively demonstrated that Aristotle's formulation does not imply that generals were regularly elected in the seventh prytany. The timing depended upon the *prytaneis* securing εὐσημία and in 188/7 B.C., for instance, the text of *IG* II^2 892, lines 5–6 reveals that the [ἐκκλησία] ἀρχαιρεσίαι κατὰ τὴν μαντ[είαν τοῦ θεοῦ] took place on the twenty-ninth day of the tenth prytany = twenty-ninth day of Mounychion, in the period of the twelve tribes. This was roughly equivalent to the ninth prytany in Aristotle's day. It was a well-known feature of the election of the ten generals, often discussed by modern scholars,[162] that each new board did not assume its duties until after the beginning of the military campaigning season.

Such flexibility in the timing of the election makes it impossible for us to fix the date of the assembly meeting at which the first board of ten men in the grain-tax law were chosen, or indeed exactly when their successors were elected. We can conclude, however, that they could not have been chosen earlier than the seventh prytany of the archonship of Sokratides, roughly March/Elaphebolion. Holding the election in the spring would ensure that the ten men were in place in time to begin overseeing the tax-grain from the islands when it began to come into the Aiakeion after each year's harvest, in other words, probably after May/June. As noted below, pages 70–73, 77, their most active duties would normally extend from election day until the late winter of the following Julian year.

In addition to considerable flexibility in the timing of the meeting when the generals were elected, Aristotle, *Ath* 44.4, indicates that the *ekklesia* was also not tied down as to the procedure of the election: ποιοῦσι δὲ καὶ ἀρχαιρεσίας στρατηγῶν . . . ἐν τῇ ἐκκλησίᾳ, καθ' ὅ τι ἂν τῷ δήμῳ δοκῇ ("they also hold elections of the generals . . . in the Assembly in accordance with whatever the people decide"). I am not concerned here with the lively debates about tribal representation on the board of generals which discussion of this passage has provoked.[163] What we should note here is that the same kind of procedural freedom that the *ekklesia* enjoyed in electing the *strategoi* could conceivably have been extended to the way they chose the ten men.

In studying Aristotle, *Ath* 44.4 in the context of other evidence for Athenian electoral procedure in the Classical period, Piérart (1974, pages 133–136) presented an instructive tabular analysis of the formulas used in some fifty Athenian decrees that called for the election or appointment of magistrates, ambassadors, commissioners, and the like. It is striking how closely Agyrrhios' law conforms to the order and the content of the several components of this formula as Piérart has extracted it from the data. I present the evidence in parallel columns.

PIÉRART	GRAIN-TAX LAW
1. La mention du corps chargé de l'election: ἑλέσθαι ou χειροτονῆσαι τὸν δῆμον ou τὴν βουλήν	αἱρείσθω ὁ δῆμος
2. Le nombre de personnes à élire	δέκα
3. La mention de leur titre (le plus souvent, on a simplement ἄνδρα, ἄνδρας)	ἄνδρας
4. Les conditions qu'elles doivent remplir: âge requis, origine (ἐξ Ἀθηναίων ἁπάντων, ἐκ βουλῆς, ἐξ Ἀρεοπαγιτῶν, etc.)	ἐξ Ἀθηναίων ἁπάντων
5. Une formule réclamant l'urgence de l'opération (αὐτίκα μάλα, ἤδη)	ἐν τῆι [ἐκ]κλησίαι, ὅτανπερ τοὺς στρατηγοὺς α[ἱρῶ]νται
6. Une proposition relative (οἵτινες . . .), participiale ou apposée précisant la mission à accomplir	οἵτινες ἐπιμελήσονται τοῦ σί[τ]ο

[161] Aristotle, *Ath* 44.4; Rhodes 1981, pp. 535–537; Pritchett 1992, p. 59, note 62.

[162] E.g., Pritchett 1940; Fornara 1971, pp. 40–41.

[163] See, e.g., Fornara 1971, pp. 1–10.

LINES 39–41

οἵτινες ἐπιμελήσονται τοῦ σί[τ]-
ο· οὗτοι δὲ ἀποστησάμενοι τὸν σῖτον̣ κ[α]-
τὰ τὰ γεγραμμένα

. . . who are to take care of the grain. After these men have seen to the weighing of the grain in accordance with the written instructions. . .

The call for election ἐξ Ἀθηναίων ἁπάντων, not for sortition, and the inferred likelihood of reelection are both in keeping with the duties of the ten men as prescribed in these lines. The law made demands upon these new officials which the Assembly may have hesitated to entrust to men selected at random. First, their final prescribed act each year was to carry up to the Pnyx from the Agora what probably amounted to a considerable sum of cash produced by the sale of the people's grain, lines 51–55. Possession of sacks or chests of silver coins clearly presented severe risks of peculation at any level of the administration. Even though the ten men had to present their accounts in the assembly, lines 52–53, and may have been subjected to audit and liable to fines or other penalties for misconduct,[164] the lawgiver wanted to give the Assembly the opportunity in advance to elect men in whose integrity it at least had some confidence.

The *polis* demanded more than honesty from the ten men, since they needed the experience and the expertise to carry out fairly intricate operations, interacting with entrepreneurial tax-farmers and grain merchants across probably a wide spectrum of knowhow, ingenuity, and cunning. They had to know enough about wheat and barley to be able to exercise quality control, to test for dampness, to recognize the presence of darnel seeds, other impurities, and possibly hidden objects intended to defraud them by increasing the weight of the grain. They had to know how to supervise the nuances of weighing and measuring the grain in the prescribed manner. They had to be familiar with the special hazards of storing a large quantity of wheat and barley in a makeshift granary for at least three or four months in the rainy winter. The people's *sitos* was under their care for this period of time and they had to take proper precautions against mildew, rot, combustion, vermin, thieves, and other enemies.[165] The officials in charge of the granaries of Philip V's garrisons were instructed ἐπισκοπείτωσαν δὲ καὶ τὰ σιτοβόλεια τῆς μὲν θερινῆς ἑξαμήνου καθ' ὃν ἂν καιρὸν ὄμβρος γένηται, τῆς δὲ χειμερινῆς κατὰ δεχήμερον καὶ ἐάν τι ῥεῦμα γεγονὸς ἦι εἰς τὸν σῖτον ἐπισκευαζέ[τ]ωσαν παραχρῆμα ("let them carefully inspect the granaries on the one hand during the six months of summer at whatever time a rainstorm might take place and on the other hand during the winter six months every ten days and if any stream of water should reach the grain, they are to make repairs immediately"), lines 21–26 of the *diagramma* cited above, page 56.

The ten men had to be numerate, literate, and intelligent enough to take over from the *poletai* and the *boule* a written catalogue of the names of the *priamenoi* (including any possible *symmoriai*) together with the numbers of *merides* these tax-farmers had in each instance contracted to deliver, probably the names of the guarantors as well. They had to use this catalogue as the basis of their own records to keep track of the grain when it was brought up from the Peiraieus to the Aiakeion. This would have been no easy task, for carts and pack animals carrying wheat and barley from several different *priamenoi* may not have reached the Aiakeion in an orderly sequence. Pack animals and carts probably had to be unloaded promptly and their contents kept separate to prevent mixing them with *merides* from other *priamenoi*. A further complication that the ten men had to note in their records is that each *priamenos* had to complete the weighing of his *merides* of grain in the Aiakeion

[164] The law does not explicitly address the topics of *dokimasia* or *euthynai* for these newly created magistrates, nor does it say anything about fines or other penalties if they failed to carry out their instructions.

[165] For discussion of some of these hazards, see Gallant 1991, pp. 97–98, 179–181. See also Migeotte, forthcoming, "Ventes," with his timely observation, "Sur les greniers des cités grecques je ne connais pas d'étude systématique, comparable par exemple à celle de G. Rickman, *Roman Granaries and Store Buildings* (Cambridge, 1971)." An excellent topic for a Ph.D. dissertation.

within thirty days of bringing it up to the city, lines 16–19. A dated, daily log of all these deliveries would seem to have been necessary.

Not only did the ten men have to keep this running written tally of the grain that came into the Aiakeion, but they were also required by the law to take action when *merides* of wheat and barley did not arrive on time: οἱ δὲ αἱρεθέντες ὑπὸ τοῦ δήμου ἐπιμελούσθων ὅπως ἂν κομίζηται ὁ σῖτος ἐν τῶι χρόνωι τῶι εἰρημένωι ("let those elected by the Assembly see to it that the grain is conveyed in the stated period of time"), lines 48–51. In the absence of specific instruction as to what procedure they were required to follow, we might guess that if the ten men found that *sitos* on their list had not arrived before the deadline of Maimakterion, it was their responsibility to report to the *poletai* and the *boule* the name of the defaulting *priamenos*, the number of *merides*, and the names of his guarantors.

The amount of bookkeeping involved in this operation was clearly considerable. When all the wheat and barley had been delivered to the Aiakeion, the ten men probably had to file their records of *priamenoi*, *merides*, delivery dates, and so forth with the *boule* and the *poletai* as verification that the tax-farmers had fulfilled their contract obligations. But the ten men had further records to keep when they sold the grain in the Agora on a day and at a price set by the assembly. These accounts of the proceeds of the sale, τὰ ἐκ τõ σίτο γενόμενα, line 55, they had to present to the *ekklesia*, λογισάσθω[ν] ἐν τῶι δήμωι, lines 52–53. Totals in these accounts had to square with the amount of grain delivered to the Aiakeion and the sums in the sacks of silver coins that they had to carry into the Pnyx. Migeotte (forthcoming, "Ventes") has a useful collection of testimonia on the written records compiled by public officials in charge of grain.

The law is silent about remuneration (μισθός) for the ten men. Probably, like most annual Athenian magistrates in the first half of the 4th century B.C., they served without pay.[166]

LINES 41–44

πωλόντων ἐν τῆι ἀγ[ορ]-
ᾶι, ὅταν τῶι δήμωι δοκῆι· πωλε̃ν δὲ μὴ ἐ[ξε]-
ῖναι ἐπιψηφίσαι πρότερον τοῦ Ἀνθεσ[τ]-
ηριῶνος μηνός

Let them sell [the grain] in the Agora at whatever time seems best to the Assembly but it is not permitted to put the sale [of the grain] to a vote before the month of Anthesterion

Finally, in addition to all these various qualifications, the ten men also had to possess the organizational skills to conduct the sale of the people's grain in the Agora. We have no details about this procedure. At the very least, however, we can imagine that the situation was full of potential for confusion, dispute, and mismanagement, as grain had to be brought out from the Aiakeion in sufficient quantities to maintain a constant supply and avoid lengthy delays.[167] A large and probably volatile crowd had to be kept in some kind of order. Wheat and barley had to be measured out from official standard containers into a myriad of different sacks, pails, and other receptacles, each of whose owners probably had firm ideas about their capacity. Perhaps identity checks had to be made of the purchasers. Suspect coins may have had to be tested,[168] and a whole range of unforeseen problems faced. It is no wonder that the *nomothetai* wanted to give the Assembly the opportunity to choose the ten men who faced this challenging assignment. They had to have the requisite organizational ability.

It is perhaps significant that the lawgiver does not place the sale of the people's grain in the grain market at Athens, ἐν τῶι σίτωι, attested in the almost contemporary law of Nikophon on

[166] On this point, see Hansen 1991, pp. 240–242.

[167] On moving the grain out of the Aiakeion for sale, see below, pages 107–108.

[168] The *dokimastes* of Nikophon's law on silver coinage, *SEG* XXVI 72, may have assisted the ten men at the sale.

silver coinage, 375/4 B.C.[169] It was here that grain in the private sector was sold on the open market. Location of the sale supervised by the ten men in the Agora is in keeping with the purpose clause at the beginning of the grain-tax law, ὅπως ἂν τῶι δήμωι σῖ[το]ς ἦι ἐν τῶι κοινῶι (lines 5–6).[170]

What exactly is meant by ἐν τῆι ἀγορᾶι? Until the time of the sale, the wheat and barley will have been stored in the Aiakeion. But the law does not explicitly place the sale at this building. Moreover, as we shall see, Herodotus 5.89.3 and *IG* I³ 426, lines 5–8, both suggest that although the Aiakeion was very close to the Agora, it was not strictly in it (below, pages 85–90, 95). It may be that the recently renovated and possibly overstuffed temenos, now serving as a public granary, lacked the space and the facilities necessary for selling large quantities of wheat and barley, probably to large numbers of buyers. For speculation about the location of the sale in the Agora, see below, Chapter III, pages 107–108.

Agyrrhios says nothing on matters that appear important to us such as restrictions on the status of the buyers of the grain (only citizens?), priorities (first come, first served?), on the quantity of grain sold to any one individual or single *oikos*, on the number of days the sale would last, on the possible resale of public grain to others, and so forth. We do not know if sales of wheat and barley continued until such time as the stock in the Aiakeion was exhausted or whether they were intentionally spread over an extended period of time with limited amounts sold each day.

Another topic on which our law is silent and comparative evidence is of limited help concerns the public announcement of the forthcoming sale of *sitos* in the Agora by the ten men. In the interest of equity, it would seem necessary at the least for the date(s), place, and perhaps some of the conditions of the sale to be posted in writing in a public place like the Monument of the Eponymous Heroes and possibly announced by heralds throughout the Attic demes. For this aspect of the sale of public grain, see Migeotte, forthcoming, "Ventes."

All of these topics, and maybe more that we cannot imagine, could have been discussed and voted upon at meetings of the *ekklesia*, perhaps at the very session when the date for the sale was set. In Aristotle's day the topic περὶ σίτου was a fixed agenda item at each κυρία ἐκκλησία, *Ath* 43.4, but the clause ὅταν τῶι δήμωι δοκῆι, line 42, in our law suggests that the *ekklesia* had considerable freedom as to when (and perhaps how) it implemented the provision regarding the sale of the people's grain.

Elsewhere some of these rules were spelled out in the published texts of the regulations governing the sale of public grain: for example, at Koroneia, grain was sold to "all" and 30 *kophinoi* were sold each day, *SEG* XLIII 205, lines 21–22; cf. Migeotte (forthcoming, "Ventes"), who suggests that all sales of this sort were made only to citizens. In large *poleis* like Athens this would have required some form of identification, perhaps by means of a dikast's *pinakion*.

Agyrrhios and the *nomothetai* did, however, impose one temporal restriction on the *ekklesia:* πωλε̃ν δὲ μὴ ἐ[ξε]ῖναι ἐπιψηφίσαι πρότερον τοῦ Ἀνθεσ[τ]ηριῶνος μηνός, lines 42–44. I take this to mean that not even a motion to sell the grain could be introduced into the Assembly earlier than the first day of Anthesterion. The use of the festival calendar to mark the *terminus* is striking. Payments of taxes in cash were almost all governed by schedules fixed to the prytany calendar. Deadlines for the delivery of the *dodekate* on the *sitos* from the islands, however, seem rather to interact with the end of the sailing season,[171] with the harvest, and with the time of year when supplies of grain in Athens were probably at their lowest level. The decree governing the public sale of grain bought by the *polis* of Koroneia regulates both its purchase and its sale in terms of

[169] See the commentary of Stroud (1974, p. 180). We do not know the location of the grain market in Athens, but Nikophon's law may imply that it was not in the Agora; see note 138 and Rhodes 1981, p. 579.

[170] We hear elsewhere of sales and distributions of grain and flour in Athens at the Pompeion and the Odeion, [Demosthenes] 34.37. Migeotte, forthcoming, "Ventes" has a valuable collection of evidence for the sales of public grain in the Greek *poleis*.

[171] On the sailing season, see Casson 1971, pp. 270–273; Reger 1993, p. 304.

the festival/lunar calendar, *SEG* XLIII 205, lines 9–10, 18–19, πρὸ τᾶς τριακάδος τῶ Ἑρμαίω μεινὸς κατὰ θεόν ("before the 30th of the month Hermaios according to the goddess [moon]"), and 19–20; similarly *SEG* XXXVI 788, line 8, Samothrace.

In our surviving records, no meetings of the *ekklesia* are apparently attested for the first day of Anthesterion, nor indeed for the *noumenia* of any other month in the Athenian calendar. Anthesterion 2–4 were monthly festival days, as were the 6th–8th. One decree of uncertain date was passed on the 8th (*SEG* XVI 97; XXXVIII 108), but otherwise the first half of this month, including the three days of the Anthesteria itself, 11th–13th, is so far devoid of securely attested meetings of the *ekklesia* in this period which do not depend upon restored calendar equations.[172] Our documentation is not full enough, however, to push any conjectural first meeting of the *ekklesia* in Anthesterion as late as the middle of the month.

For our purposes the important point to note is that the people's grain was held in storage until at least the winter month of Anthesterion, roughly equivalent to February/March.[173] It seems evident that the intent of the prohibition about introducing a motion in the *ekklesia* to sell the grain any earlier than Anthesterion was to make a certain quantity of wheat and barley available to the public at the time of year when maritime activity was normally reduced, existing supplies of grain were often low, and prices presumably approaching their peak. For grain sales elsewhere at this time of year before the new harvest, see Gauthier 1987.

LINES 44–51

ὁ δὲ δῆμος ταξάτω τὴν τ[ι]-
μὴν τῶν πυρῶν καὶ τῶν κριθῶν ὁπόσου χ[ρ]-
ὴ πωλε͂ν τοὺς αἱρεθέντας

Let the Assembly set the price at which the elected men must sell the wheat and barley

The law instructs the Assembly to set not only the date for the sale but also the prices of wheat and barley. For the formulation, cf. Aristotle, *Ath* 39.3, ἥντιν' ἂν οὗτοι τάξωσιν τιμήν ("at whatever price these men set"); *Syll.*^{3} 976, lines 25–26, διδόντες τῆι θεῶι τιμὴν μὴ ἐλάσσονα ἧς πρότερον ὁ δῆμος τέταχεν ("giving to the goddess a price no less than that which the Assembly has previously set"), Samos. The accounts of the *epistatai* of Eleusis, 329/8 B.C., provide a close parallel for both the wording and the procedure at Athens, κεφάλαιον τιμῆς κριθῶν . . . πραθεισῶν ἐκ τριῶν δραχμῶν τὸν μέδιμνον ἕκαστον ὡς ὁ δῆμος ἔταξεν ("total of the price of barley . . . sold at three drachmai per *medimnos* as set by the Assembly"), *IG* II^2 1672, lines 282–283; same formula used for wheat in lines 286–287. Cf. τιμᾶς ἅστινός κα ὁ δᾶμος ἐπιχειροτονείσει ("at whatever price the Assembly will decide by vote"), *SEG* XLIII 205, lines 22–23, Koroneia. Migeotte (forthcoming, "Aspects," note 30) aptly cites Pollux 7.14, οὔτε τιμῆς τεταγμένης πωλοῦσιν, ἀλλ' ὡς ἂν δύνωνται πλειστηριάσαντες πλείστου ἀπέδοντο ("they do not sell it at a fixed price but they sold it at the highest level, raising the price as much as they could") = Lysias Fr. 20, Gernet. Cf. Migeotte 1997, page 38 and note 26.

For students of the ancient economy, lines 44–46 of Agyrrhios' law provide firmly documented and precisely dated evidence for price-fixing of public grain by the *polis*.[174] In having the *ekklesia*

[172] Mikalson 1975, pp. 111–121. For decrees containing restored dates of Anthesterion 8 and 9, respectively, see *SEG* XXXIV 72 and XXXIX 101.

[173] At Koroneia grain purchased by the *polis* remained in storage for almost a year before it was sold publicly; see *SEG* XLIII 205, with the discussions of Migeotte 1991, p. 27; forthcoming, "Ventes."

[174] From the massive bibliography on this topic, particularly helpful are Ampolo 1986; Fantasia 1987, pp. 111–117; Garnsey 1988, pp. 69–86; Gargola 1992; and especially Migeotte (forthcoming, "Aspects" and "Ventes"), who collects excellent parallels and presents persuasive conclusions. Our law now brings helpful instruction to views of the ancient Athenian economy, such as Hopper's (1979, p. 59): "It is odd that there is no suggestion that the State should step in, and by the purchase and storage of

set the price, the lawgiver presumably hoped to avoid the kind of manipulation of the market by the grain dealers outlined in Lysias 22.15 (387/6 B.C.): ὅταν γὰρ μάλιστα σίτου τυγχάνητε δεόμενοι, συναρπάζουσιν οὗτοι καὶ οὐκ ἐθέλουσι πωλεῖν, ἵνα μὴ περὶ τῆς τιμῆς διαφερώμεθα, ἀλλ' ἀγαπῶμεν ἐὰν ὁποσουτινοσοῦν πριάμενοι παρ' αὐτῶν ἀπέλθωμεν ("For whenever you happen to be most in need of grain, these men grab it up and are unwilling to sell in order that we won't haggle over the price, but we feel good if we come away from them having made a purchase at any price at all"). In this speech Lysias twice (17 and 21) accuses the *sitopolai* of conspiring against the ἔμποροι and the εἰσπλέοντες.

We cannot know the sale price of a *medimnos* of the people's wheat or barley from the islands in 374/3 B.C. We have virtually no evidence of closely contemporary prices of wheat and barley on the open market at Athens, and the fluctuations attested in Lysias 22 should warn us against putting much faith in anything like "a standard price."[175] In the period 414–328 B.C., from our very imperfect record, a *medimnos* of wheat appears to have fetched as little as five drachmai and as much as sixteen,[176] a *medimnos* of barley from two to six-plus drachmai.[177] But these prices cannot be regarded as reliable guides, since we do not know in each case the time of year when they were in effect and for how long. Following the same procedure as that provided for in the grain-tax law, the Athenian Assembly in 329/8 B.C. set the prices at which the Eleusinian *epistatai* sold the surplus wheat and barley from the *aparchai* to Demeter and Kore at six and three drachmai per *medimnos*, respectively, *IG* II^2 1672, lines 282–283, 287.

We have even less evidence on which to speculate about the factors that went into setting the price of wheat and barley in the Assembly in 374/3 B.C. and later. On a purely physical level, for all we know, there could have been significant differences in quality between, say, the wheat produced by the fertile, volcanic soil of Lemnos, and that from the rockier Skyros.[178] Quality

---

corn (in the manner of Gaius Gracchus at Rome), and its release in times of crisis, get round this problem. That the State did not do so was not so much a matter of dislike of intervention in this sector of the citizen's life, as of a lack, in the primitive organization of Athens, of the necessary financial and administrative machinery." Cf. the notion of the weakness of the *polis* expressed by Gallant (1991, p. 26).

[175] Attempts to postulate a "normal price" or a "regular" price for grain, based upon a mistranslation of ἡ καθεστηκυῖα τιμή, go back at least as far as Boeckh 1886, I, p. 118, note c; II, p. 26, note 63, and extend at least to Millett in Cartledge and Todd 1990, pp. 192–193. Migeotte, forthcoming, "Aspects," note 28 gives a helpful list of followers of this view. Ampolo (1986, pp. 146–147) and Migeotte (*ibid.*) have effectively shown that this expression means rather a fixed price or an established price. Cf. also Reger 1993, pp. 312–314: "There is no single 'normal' price: grain prices are highly time-dependent." For the dangers of postulating an average price of grain at Rome, see Rickman 1980, pp. 239–240. Even in 3rd-century B.C. Delos, where evidence for grain prices is more plentiful, there are fluctuations as great as +51% and –25%. See the useful study by Reger (1993, pp. 304–317), who also prudently warns with regard to price ratios between wheat and barley, "The striking difference between this figure [Delos] and those for Athens and Egypt should discourage facile assumptions about the universal applicability of price ratios derived from one place and one time, and indeed it is the differences, not the similarities, that require emphasis" (p. 307). Also, "In any given year, indeed at any given moment, a congeries of factors worked to establish prices for grains. Data from Egypt suggest the extraordinary daily volubility of prices and the great difference between pre- and post-harvest prices" (p. 308).

Caution is probably also advisable in arguing for a norm in the relative prices of wheat and barley, i.e., a *medimnos* of the former being worth roughly twice the equivalent amount of the latter; see Knoepfler 1988, pp. 292–293. For ἡ καταξία τιμή, which I would translate as a "fair price," see *IG* V.1.1379, line 24, with the discussion of Migeotte (forthcoming, "Ventes").

[176] Wheat: 5 drachmai, [Demosthenes] 34.39; *IG* II^2 360, lines 9–10; 16 drachmai, [Demosthenes] 34.39; Pritchett 1956a, p. 197.

[177] Barley: 2 drachmai, Epiktetos Fr. 11; 6-plus drachmai, [Demosthenes] 42.20, 31. I am indebted to W. R. Loomis for helpful discussion on the prices in these two notes.

[178] This may have been the case in later times; see Aelian's characterization of Skyros as λυπρὰ καὶ ἄγονος καὶ ἀνθρώπων χηρεύουσα ὡς τὰ πολλά, *On the Nature of Animals* 4.59; echoed by Philippson (1901, p. 121), "die unfruchtbarste aller grösseren griechischen Inseln," and others, e.g., Jardé (1925, p. 179); W. Günther (in Lauffer 1989, p. 627). See, however, the qualifications to this somber picture added by the travelers cited above, note 73 and by Graindor (1906, pp. 13–15). An invaluable resource on Skyros, ancient and modern, is Antoniades 1977, which assembles and translates all the travelers' descriptions of the island, with excellent

undoubtedly varied from place to place and from year to year. The ten men, through whose hands all of this grain had to pass before the meeting of the *ekklesia*, would have been in a position to alert their fellow citizens to physical factors like this which might have led to differentiation in the price. It seems to me unlikely that the opportunity to sell a higher-quality product at an increased price would have been passed up simply by pouring all the grain from all three islands into one common bin and selling it as a uniform batch. The fact that the Assembly basically set one price for the wheat sold by the Eleusinian *epistatai* in 329/8 B.C. and one price for the barley need not rule out differences in prices in 374/3 B.C. The grain that came into Eleusis as *aparchai* in 329/8 B.C. (*IG* II^2 1672) arrived in much smaller individual units than that envisaged by the grain-tax law and from seventeen different points of origin, including the ten nonterritorial Athenian tribes. Price differentiation on the basis of quality would have been very difficult under such conditions, even though a certain quantity of wheat was in fact sold for five drachmai per *medimnos* rather than six; see lines 287–288.

Political and economic factors probably helped shape the debate in the *ekklesia* as to when and for how much the ten men should sell the people's grain. Here we must be careful not to be drawn into the simplistic trap of trying to identify "*the* purpose of the law" and rigidly inferring from it the likely tenor and outcome of the debate. It is also probably a mistake to regard public sales of grain by a *polis* as always motivated exclusively by the desire either to provide food at lower than market prices or to turn a civic profit. We must be prepared, I think, to entertain a more nuanced political approach. For instance, some speakers focusing on the introductory purpose clause of the law, ὅπως ἂν τῶι δήμωι σῖ[το]ς ἦι ἐν τῶι κοινῶι, lines 5–6, might have pressed for a very low price, well below the current rates for wheat and barley on the open market. After all, this was the people's grain, intended for the *demos*. It originated in a tax voted for and imposed by the *polis*. Individual members of the *demos* ought to reap the benefits of this tax, which cost them little, by being able to buy cheaply a certain quantity of wheat and barley that had been sequestered from the fluctuating prices on the open market, especially now at the end of winter. For the view that in 329/8 B.C. the *demos* had a personal interest in setting a low rate, see Garnsey 1988, pages 154–162; Reger 1993, pages 306–307, 312.

On the other hand, it surely is not too fanciful to imagine that other speakers might have focused on another clause in the law, ἔστω στρατι[ω]τικ⟨ὰ⟩ τὰ ἐκ τõ σίτο γενόμενα, lines 54–55. Without invoking civic ideology, the struggle between mass and elite, and the like, we might imagine that these speakers sought to persuade their fellow Athenians that it would be foolish to pass up or diminish this excellent opportunity to increase the income of the military fund and strengthen the *polis*. After all, it did not cost the *demos* much to fill the Aiakeion with tax-grain. Why not soften the blow of other taxes, especially the *eisphora*, by realizing as much as possible from the sale of the wheat and barley from the islands? Set the price just a little bit below the current prices on the open market. Athenian buyers would thereby beat the going rate while at the same time contributing patriotically to the revenues of the *polis*.

Readers can easily reconstruct other possible lines of argument. My only purpose in speculating so wildly on the basis of virtually no evidence has simply been to underline the probable complexity of the issues of the timing and the price of the sale of this grain as they were debated in the *ekklesia*. The combination of taxes, prices, food, and the military treasury is not likely to have led to quick and easy resolution or to unanimity of purpose. Nor must we be surprised if public debate on the topics of finance, economy, and revenue reached a fairly high level of sophistication.[179] For

---

bibliography and valuable introductions and commentary. On the fertility of Lemnos, see now briefly Salomon 1997, pp. 175–176.

[179] I am fully persuaded by the perceptive warning of Hanson (1995, p. 322): "Do not be carried away by notions of the 'primitiveness' of the Greek economy to such an extent as to believe the ancients had no idea of profit and loss, income and outlay. Economically the Greeks may not have acted in a purely capitalistic mode in the modern sense. . . . But the farmer-citizens of the *polis* were conscious of how a market worked, of debits, incomes, and expenditure." See also the demonstrations

debate in the *ekklesia* on the food supply, see Aristotle, *Rhet.* 1.1360a–b12–17; Isager and Hansen 1975, pages 27–29.

We cannot know the magnitude of the sums of money that might have been at stake in these deliberations. Potentially, however, they are unlikely to have been insignificant. Taking our very crude, minimum estimate of 6,200 *medimnoi* of wheat and 24,800 of barley extrapolated from the *aparchai* accounts of 329/8 B.C., above, page 41, and using *exempli gratia* 6 and 3 drachmai per *medimnos* for the prices of wheat and barley, respectively, we could guess that the public sale might have generated 6,200 × 6 = 37,200 dr. + 24,800 × 3 = 74,400 dr., for a total of 111,600 drachmai, or roughly $18\frac{1}{2}$ talents. This is probably a very minimum figure that would not have made a huge dent in Athenian military costs, but even this small sum would have been enough to keep eighteen triremes on active duty for a month.[180]

We might also wonder whether in voting for the allocation to the military fund of revenue from a tax in grain some Athenians might have expected to help pay for naval convoys or squadrons to protect the grain ships. See Lysias 19.50–52; [Demosthenes] 50.17–23, ἐπὶ τὴν παραπομπὴν τοῦ σίτου; Gernet 1909, pages 347–364; Jardé 1925, pages 197–198; Isager and Hansen 1975, page 57; Hopper 1979, page 83; Velissaropoulos 1980, pages 135–136.

Before leaving this provision of the law, I note one aspect that is not very speculative or controversial. It is so characteristic of democratic Athens that Agyrrhios made no attempt to legislate in advance regarding the price of the people's grain. It was not something to be fixed by a board of *sitophylakes* or a panel of "experts." The *demos* had the right, indeed the obligation, to settle by debate the price of its own grain. This provision appears exactly to fit the historical and political situation in 374 B.C., for we have no reason to believe that this was a time of crisis, *sitodeia*, or the like, in which the *polis* intervened on an *ad hoc* basis in the setting of grain prices. Agyrrhios carried a *nomos* with long-range implications, not a *psephisma* aimed at fixing a one-time emergency. This price-fixing by the *ekklesia* was an integral part of an ongoing plan similar in a way to the role of the Athenian Assembly in annually setting the price at which the wheat and barley produced by the annual Eleusinian *aparche* would be sold (*IG* II^2 1672, lines 282–283, 286–287). These two examples of long-term price-fixing by the Athenian *ekklesia*[181] stand out against the numerous examples of crisis-control by the *polis* ably collected and illuminated by Migeotte 1997. Arising out of annual taxes or quotas imposed on production of grain in Attica and Athenian possessions abroad, they suggest that price-fixing by the *polis* at a deeper structural level might have been in place earlier and on a more regular basis than some scholars have so far been prepared to believe. For the *polis* as an agent in the distribution of grain, see Gernet 1909, page 375; Jardé 1925, pages 198–200; Garnsey 1988, pages 74–86.

For Maimakterion in line 48 marking the beginning of winter and the end of the sailing season, see above, pages 72–73.

---

by W. E. Thompson (1978; 1982) and Cohen (1992) of the inadequacy of a "primitivist" or "substantivist" approach to Athenian finance in the 4th century B.C. Migeotte (1996, pp. 82–89) has an excellent discussion of the involvement of the Assembly in the Greek *poleis* in public finance and the grain supply. Cf. also Fantasia 1987, pp. 111–117.

[180] For this computation, see Thucydides 6.8.1; Casson 1971, p. 302, note 7. For the possibility that the estimate of one talent per month to operate a trireme might be too high, see Wilson 1970, pp. 304–305; Wallinga 1993, pp. 170–178.

[181] Kevin Clinton reminds me, however, of the absence of any provision for price-fixing of the *aparchai* by the *ekklesia* in the 5th-century B.C. First-Fruits decree, *IG* I^3 78. In lines 40–43 the *hieropoioi* are merely instructed to sell the wheat and barley and the *demos* decides what dedications are to be made from the proceeds.

LINES 51–55

ἐπειδὰν δὲ ἀποδῶντα-
ι οἱ αἱρεθέντες τὸν σῖτον, λογισάσθω[ν]
ἐν τῶι δήμωι καὶ τὰ χρήματα ἡκόντων φ[έ]-
ροντες εἰς τὸν δῆμον καὶ ἔστω στρατι[ω]-
τικ⟨ὰ⟩ τὰ ἐκ τõ σίτο γενόμενα

Whenever the men who have been elected have sold the grain, let them render their accounts in the Assembly and let them come into the Assembly carrying the money and let the amount that is realized from the sale of the grain be assigned to the military fund

Agyrrhios here lays down the final instructions for the ten men. For this formulation see the 5th-century B.C. Athenian decree on the Eleusinian First-Fruits, *IG* I^3 78, lines 40–41, τὰς δὲ ἄλλας κριθὰς καὶ πυρὸς ἀπ|οδομένος τὸς *h*ιεροποιός ("when the hieropoioi have sold the rest of the barley and the wheat"), and, once again, the accounts of the Eleusinian *epistatai* for 329/8 B.C., *IG* II^2 1672, lines 281 and 285, ἀπεδόμεθα μεδίμνους... For an excellent discussion of the use of ἀποδίδωμι as the aorist of πωλέω, see Chantraine 1940, pages 18–21.

The ten men are to present their accounts of the public sale of the people's grain in the Assembly, λογισάσθω[ν] ἐν τῶι δήμωι. The verb used here in the middle without an object and with magistrates as its subject I take to be the equivalent of the formulation *h*οι αἰεὶ ταμίαι ... λόγον διδόντον τõν τε ὄντον χρεμάτον ... πρὸς τὸς λογιστάς ("let the treasurers in office at the time render an account of the existing funds ... before the auditors"), *IG* I^3 52, lines 24–27; and καὶ λογιστὰς δέκα καὶ συνηγόρους τούτοις δέκα, πρὸς οὓς ἅπαντας ἀνάγκη τοὺς τὰς ἀρχὰς ἄρξ[α]ντας λόγον ἀπενεγκεῖν. οὗτοι γάρ εἰσι μόνοι ⟨οἱ⟩ τοῖς ὑπευθύνοις λογιζόμενοι ("and there are ten auditors and ten assistants for them before whom it is mandatory for all those who have served as magistrates to render their accounts. For these are the only ones who audit the accounts of men who are subject to *euthynai*"), Aristotle, *Ath* 54.2. It is possible that the *ekklesia* was assisted in its scrutiny of the accounts of the ten men by *logistai*, although they are not mentioned in the text, nor is anything said about εὔθυναι. This inscription cannot help to bridge the gap that exists in our record between the thirty *logistai* attested in the 5th century B.C. and the ten of the 4th, represented inter alia by the two passages just quoted. For good discussions of these officials, see Piérart 1971, pages 562–573; Rhodes 1972, page 111; 1981, pages 560–561, 597–598.

The strikingly graphic instructions for the ten men, τὰ χρήματα ἡκόντων φ[έ]ροντες εἰς τὸν δῆμον (lines 53–54), are charming, almost Aristophanic. They give us a glimpse of what could have been staged as a grand ceremony, perhaps witnessed by a large crowd:[182] the ten men and their attendants carrying bags or boxes of coins moving in a stately procession, possibly with an armed escort, from the southwest corner of the Agora and their headquarters at the Aiakeion, up along the road in the narrow hollow between the Areiopagos and the Hill of the Nymphs, turning into a crowded Pnyx where the *demos* sits ready to receive the proceeds from the sale of its grain and to learn how much it will turn over to the military fund. Probably waiting there with the *ekklesia* and the *boule* was the ταμίας τῶν στρατιωτικῶν.[183] Their route, conveniently, would have taken them right through the "home district" of the lawgiver Agyrrhios, the deme of Kollytos; see above, page 17.

[182] One is reminded of the annual procession in the theater of Dionysos when the allies in the 5th century B.C. brought their tribute to Athens; see Isokrates 8.82; Smarczyk 1990, pp. 155–167. L. Migeotte points out to me that the ten men would certainly require assistance, for $18\frac{1}{2}$ talents represent almost 500 kg. of metal. For the wording, Adele Scafuro notes [Dem.] 56.18, ἐξὸν αὐτῷ ἐπὶ τὸ δικαστήριον ἥκειν φέροντα τὸ ἀργύριον.

[183] For his role in the farming out of taxes, see above, p. 29. Although Agyrrhios legislated that the proceeds of the *dodekate* would eventually be allocated to the fund administered by this official, the ten men may have handed the cash over to the *apodektai* and the *boule*, who in due course made the allocation to the treasurer of the military fund.

Although Agyrrhios left to the *ekklesia* the election of the ten men and the decision as to when and at what price they were to sell the people's grain, he preempted any possible debate on the disposition of the funds produced by that sale. Written explicitly into the law is the provision ἔστω στρατι[ω]τικ⟨ὰ⟩ τὰ ἐκ τõ σίτο γενόμενα, lines 54–55. Thus the $8\frac{1}{3}$% tax on the grain from the islands and the *pentekoste* eventually produced in Athens revenue that was earmarked in advance for this specific purpose. We have seen above, pages 74–76, the possibility that the potential sum of money was not enormous, but substantial.

To my knowledge, this passage provides the earliest evidence at Athens for the military fund, τὰ στρατιωτικά. It does not in fact predate by much our previously earliest surviving testimony, the deposition of Timotheos as *strategos* in the archonship of Asteios, 373/2 B.C. In rendering his accounts after he was deposed and prior to his trial in Maimakterion of this year, Timotheos claimed that he had distributed seven minai for his own ships and one thousand drachmai for the Boiotian ships ἐκ τῶν στρατιωτικῶν χρημάτων ([Demosthenes] 49.12 and 16). There is nothing in the grain-tax law or this oration to suggest that the fund was of very recent origin, and no hint as to whether it was at this time a regular annual recipient in the *merismos*. Cawkwell's view that the formation of the *stratiotika* went back to the inception of the Second Athenian Confederacy in 378 B.C. thus receives some slight support from our inscription.[184] We can add now one certain annual source of revenue for the fund after 374/3 B.C.

LINES 55–59

τὴν δὲ προ[κ]-
αταβολὴν τὴν ἐκ τῶν νήσων μερίσαι το[ὺ]-
ς ἀποδέκτας καὶ τῆς πεντηκοστῆς, ὅσο[ν]-
περ πέρυσιν ⟨η⟩ὗρεν ἐκ τοῖν δυοῖν δεκάτ[.]|ιν

Let the Receivers allocate the down payment from the islands and of the 2% tax exactly as much as was fetched last year from the two tenths

At the end of line 55, the lawgiver, rather abruptly, shifts his attention from the ten men to the *apodektai*. His syntax and prose style shift, too, as he abandons the imperative mood that dominated lines 36–54 to employ the accusative and infinitive construction τὴν δὲ προ[κ]αταβολὴν τὴν ἐκ τῶν νήσων μερίσαι το[ὺ]ς ἀποδέκτας, lines 55–57. Two more infinitives, εἶναι, line 59, and ⟨ἀ⟩φαιρεῖν, line 60, extend this construction through to the end of the law.

To the modern reader, a προκαταβολή (down payment) from the islands might seem initially surprising after line 27, which explicitly stated, in the context of the *dodekate*, προκαταβολὴν οὐ θήσει ὁ πριάμε[ν]ο[ς]. Moreover, to my knowledge, προκαταβολαί in public transactions were always made in cash.[185] That we are certainly dealing in lines 55–61 with a cash down payment from the islands is also clear from what the *apodektai* are directed to do with it: μερίσαι.[186] Not only are the *priamenoi* of the *dodekate*, however, excused from a *prokatabole* in our law, but with the exception of payment of the *eponia* and *kerykeia*, lines 28–29, there is no indication of cash passing from the tax-farmers to the *polis* in all the lengthy description of the $8\frac{1}{3}$% tax on the *sitos* from the islands, lines 6–55.

These considerations raise the possibility that the προκαταβολὴ ἐκ τῶν νήσων, rather abruptly introduced in lines 55–56, is somehow not an essential part of the δωδεκάτη τοῦ σίτου τῶν νήσων, as it is defined and explicated in the text of this law. We have seen that the yield of the latter

[184] Cawkwell 1962; 1963a, pp. 60–61; Rhodes 1972, pp. 105–110; 1980, pp. 309–312; Brun 1983, pp. 170–177.

[185] See above, p. 61.

[186] For the *merismos* in Athenian finance, see Rhodes 1972, pp. 98–113; Hansen 1991, pp. 262–263.

tax is expressed throughout in terms of grain, not cash. Assuming, however, that "the islands" in both of these formulations are likely to be the same, that is, Lemnos, Imbros, and Skyros, we might conclude that the cash down payment, which the *apodektai* are now instructed to disburse, originated in a different tax levied in the islands or possibly in an earlier form of the *dodekate* before it was transformed into a tax in kind. At present, I do not see clear evidence in our law to decide between these two possibilities. The abrupt and rather condensed reference simply to "the down payment from the islands" would presumably not have been ambiguous to the contemporary Athenian reader.

Reading on in line 57, we encounter another tax, καὶ τῆς πεντηκοστῆς. The genitive may indicate that we are here dealing with the down payment of the *pentekoste* or, if the genitive is partitive, that the law refers to a portion of this tax equivalent to that fetched last year from the two *tenths*, lines 57–59. This πεντηκοστή is likewise productive of a cash yield, for the *apodektai* allocate it (μερίσαι), as they do the προκαταβολή in lines 55–56. At first glance, it might be tempting to identify this πεντηκοστή, which here lacks any form of qualification, except perhaps the (for us) enigmatic ὅσο[ν]|περ clause, with the best-known 2% tax at Athens. This would be the *pentekoste* on goods coming in and out of the Peiraieus, the 2% tax in fact that Agyrrhios himself had farmed in 402/1 B.C. and Andokides in 401/0 B.C. (above, pages 19–20). On this tax, see Hansen 1991, pages 260–261.

While there might be some initial attraction to identifying τῆς πεντηκοστῆς in line 57, taken completely out of context, with this familiar 2% tax in the Peiraieus, it is not immediately obvious why Agyrrhios should be legislating about this levy for the first time in the closing section of his law on the grain-tax in the islands. It is probably more productive to look first to the context of our law for the identity of τῆς πεντηκοστῆς in line 57. This tactic leads us immediately back to the only other prior occurrence of the term in our text: lines 6–8, τὴν δωδεκάτην πωλ[εῖ]|ν τὴν ἐν Λήμνωι καὶ Ἴμβρωι καὶ Σκύρω[ι κ]|αὶ τὴν πεντηκοστὴν σίτο.

The 2% tax named in line 57 lacks the qualification σίτο of line 8, but it is suggestively linked with a "down payment from the islands" in lines 55–56. Again, we must assume, I think, that these islands are Lemnos, Imbros, and Skyros, the only such islands named elsewhere in the law. This would permit the suggestion that the 2% tax in line 57 is to be identified with its namesake in line 8. We considered above, pages 38–39, the possibility that the latter could have been a hitherto unknown tax farmed in Athens but levied on harbor activity in the islands, just like the *pentekoste* ἐν τῆι Νέαι of *IG* II^2 334 + *SEG* XVIII 13, the law on the Lesser Panathenaia, 336–334 B.C. Robert (*Hellenica* XI–XII, 1960, page 193) has shown the wide distribution of such 2% taxes on harbors throughout the Greek world.

An immediate barrier to identifying the 2% tax in line 57 of our law, however, with the πεντηκοστὴ σίτο of line 8 is raised by the fact that the former was clearly calculated in cash, whereas the law of 374/3 B.C. now mandates that the yield of the latter is to be collected in kind, in *sitos*. We saw that a similar difficulty stood in the way of identifying the source of the cash "down payment from the islands" in lines 55–56 with the *dodekate*. I suggest that the solution to both problems may be the same.

It is worth considering the following hypothesis. The reason why the *pentekoste* in line 57 is defined in terms of cash, while in line 8 it is clearly a tax in kind, is not that they are two different taxes but because Agyrrhios in 374/3 B.C. legislated this change in the status of the 2% tax. The cash proceeds of this tax mentioned in line 57 would then represent its yield prior to the present legislation. In line 8 it has already acquired its new status as a tax in kind. Such a hypothesis would also permit the suggestion that the cash "down payment from the islands" in lines 55–56 belonged to the *dodekate* on the grain in Lemnos, Imbros, and Skyros which likewise was now to be collected in wheat and barley, whereas previously it was an ad valorem tax collected in cash. This προκαταβολή, formerly required of those who farmed the $8\frac{1}{3}$% tax (οἱ πριάμενοι τὴν δωδεκάτην, lines 46–47), was now legislated out of existence by Agyrrhios—προκαταβολὴν οὐ θήσει ὁ πριάμε[ν]ο[ς], line 27.

This double hypothesis provides an answer to the question we posed above (page 16): whether the law of Agyrrhios now introduced the δωδεκάτη and the πεντηκοστή for the first time, or was its purpose to transform two previously existing taxes assessed in cash into taxes that henceforth would be collected in kind. We have also argued above that the genitive σίτο in line 8 is not to be taken closely with τὴν πεντηκοστήν as a qualifier of this tax; see pages 37–39.

Philippe Gauthier, to whom I am deeply indebted for much valuable advice on the interpretation of these lines, has suggested to me that they represent an abrupt and condensed addendum to the main body of the law. It is almost as if the lawgiver here responds to a question about how the transition is to be made from a system in which the *dodekate* and the *pentekoste* were collected in cash to one in which they will now produce *sitos*. Agyrrhios proposes a year of transition, one in which the *apodektai* are to allocate once (μερίσαι) a sum of money to the *dioikesis*, while hereafter they are not to subtract (μὴ ἀφαιρεῖν) the two *tenths* from the money that is being paid in. Gauthier aptly notes the shift in tense in the two infinitives from the single-action aorist to the continuous-action present.

It is on the *prokatabole* from the islands and on the 2% tax (whatever its origin) that the *apodektai* are now directed to focus their attention. Presumably they have, or will have, the προκαταβολή on hand as soon as these taxes are farmed and the successful bidders make their down payments. It is likely that the rest of the funds from this tax that the *apodektai* disburse originate in the cash paid to the *polis* by the tax-farmers according to a schedule of payments tied to the prytany calendar, καταβολαί (see Aristotle, *Ath* 47). The *apodektai*, for a reason that is not stated in the law, are directed to disburse out of this money a sum equivalent to ὅσο[ν]περ πέρυσιν ⟨η⟩ὗρεν ἐκ τοῖν δυοῖν δεκάτ[.]ιν.

There remains for us considerable obscurity in this passage. It probably results from the fact that the lawgiver deals in fairly compressed terms with topics and taxes that were familiar to most of his fellow citizens. For instance, he goes beyond our comprehension in speaking about revenue from "last year." We have no access to this information. Second, what is the subject of ⟨η⟩ὗρεν in line 58? Furthermore, I, for one, do not immediately recognize the two *tenths* of lines 58–60. On what were these "two tithes" imposed and where? Why are there two of them, presumably in some way closely combined? Why is the reference merely to some unexpressed total from "last year" rather than to a specific stated amount?

I cannot shed any light on the significance of πέρυσιν, that is, the archonship of Hippodamas, 375/4 B.C., in line 58. Was there something special or unusual about the two *tenths* in that year, or indeed in the archonship of Sokratides, 374/3 B.C., that called for separate treatment? Or is the lawgiver merely directing his attention to a portion of the *dodekate* and the *pentekoste* equivalent to a sum already established and known, at least to some? Compare the reference in a different context to "last year's yield" on a sea-finance loan in the contract quoted in [Demosthenes] 35.13, τοὺς τόκους ἀποδόντων τοὺς πέρυσι γραφέντας εἰς τὴν συγγραφήν ("let them pay the interest recorded in the contract last year"). Here, however, the amount has already been stated in section 10 of the speech (Cohen 1992, pages 54–55).

Candidates for the subject of ⟨η⟩ὗρεν in the relative clause in lines 57–59 include ἡ πεντηκοστή carried over from its position as a possible antecedent. This would presumably yield the translation "and of the *pentekoste* exactly as much as it fetched (or yielded) last year from the two *tenths*." This seems unsatisfactory in that it appears to mean that one tax, the *pentekoste*, produced a yield or revenue out of two other taxes, the two *dekatai*, if that is the meaning of the latter. What sense this would make is not clear to me. Syntactically also, had this been the lawgiver's intent, why did he not write ὅσο[ν]περ ἡ πεντηκοστὴ πέρυσιν ⟨η⟩ὗρεν ἐκ τοῖν δυοῖν δεκάτ[ο]ιν?

A second possibility would be to import or understand some term such as ἡ πόλις ⟨η⟩ὗρεν, "exactly as much as the *polis* realized." This seems a rather desperate solution leading to an even more compressed formulation than we already have.

Perhaps a more plausible suggestion is to take the pronoun ὅσο[ν]περ as the subject of ⟨η⟩ὗρεν, "exactly as much as last year's yield from the two *tenths*." For this sense of εὑρίσκω, which in English we might expect would require the passive voice, see Dittenberger's excellent note on *Syll.*^{3} 966, lines 36–37 (*IG* II^2 2492), τοῦ δὲ εὑρόντος ἀργυρίου, with, e.g., Herodotus 1.196.2; Aischines 1.96; Xenophon, *Memorabilia* 2.5.5; *Poroi* 4.25, ὅσον τὸ τέλος εὕρισκε τῶν ἀνδραπόδων πρὸ τῶν ἐν Δεκελείᾳ ("how much the tax on slaves fetched before the events in Dekeleia"); 4.40, ὑμεῖς δὲ ὅσα μὲν πρὸ τῆς εἰρήνης χρήματα εὕρισκε τὰ τέλη, ἀπὸ τοσούτων καὶ τὸ ἐπιὸν ἔτος διοικεῖτε τὴν πόλιν· ὅσα δ' ἂν ἐφευρίσκῃ ("You should administer the city during the coming year at the same level as what the taxes fetched before the peace. But as for anything that is fetched over and above that figure"), with the helpful commentary of Gauthier (1976, pages 157–159, 171–174). Cf. also *IG* II^2 1176, line 28, with Wilhelm 1923, pages 435–439 (I owe this reference to L. Migeotte); *IG* V.1.1379, line 24, [εἰ δέ κα μὴ εὑρί]σκ̣ει ἐ̣ν τοῖς συνέδροις ὁ σῖτος τὰν καταξίαν τιμάν ("if the grain does not fetch a fair price in the *synedria*"); *IG* VII 3073, line 38, ὅτι δ' ἂν εὕρῃ.

The *apodektai* are to allot to the *dioikesis* out of the money produced by the *prokatabole* of the *dodekate* and the *pentekoste* a sum equivalent to that fetched last year from the two *tenths*. For discussion of these *tenths*, see below pages 82–84.

LINES 59–61

τὸ μὲν νῦν εἶναι εἰς τὴν διοίκησι[ν κ]-
αὶ τὸ λοιπὸν μὴ ⟨ἀ⟩φαιρεῖν τὼ δύο δεκάτ[.]
ἐκ τῶν κατ⟨α⟩βαλλομένων χρημάτων

For the present it [the money] is to belong to the financial administration and for the future let them not take the two tenths away from the money that is being paid in

We turn now to the two final instructions to the *apodektai*. Their activity is divided into two stages, marked by τὸ μὲν νῦν . . . and [κ]αὶ τὸ λοιπὸν . . . These may be equivalent to what is to be done in the current year, and what action they are to take in subsequent years. At any rate, I take as subject of the infinitive εἶναι in line 59 an amount equivalent to last year's yield from the two *tenths*. The immediate target of these funds in the *merismos* of the *apodektai* is εἰς τὴν διοίκησι[ν]. They are to allocate or disburse this money, not to the military fund (*stratiotika*), which is to be the destination of the income from the sale of the people's grain, as we have seen (lines 53–55), but to the general fund.

Διοίκησις here presumably has its specific technical meaning, designating a general fund in the financial administration of the *polis*, which was not otherwise targeted for individual use, such as στρατιωτικά, or θεωρικά, and so forth. It was administered by the *boule*, assisted by the *apodektai;* cf. Lysias 30.22, ἡ βουλὴ ἡ ⟨ἀεὶ⟩ βουλεύουσα, ὅταν μὲν ἔχῃ ἱκανὰ χρήματα εἰς διοίκησιν, οὐδὲν ἐξαμαρτάνει ("Whenever the Council in office at a given time has adequate funds for the general administration, it does not go far wrong"), after 399 B.C.; see also Rhodes 1972, pages 106–110; 1981, pages 514–516. Its day-to-day operation, providing cash stipends for attendance of the *ekklesia*, for the *dikastai*, for the *hippeis*, and so forth, and at the same time as a recipient of cash payments from tax-farmers and all those who owed the *polis* τά τε ἱερὰ καὶ τὰ ὅσια χρήματα, is well described by Demosthenes (24.96–102, 353 B.C.). It was the topic of a famous speech by Lykourgos, περὶ τῆς διοικήσεως.[187] His own title for a time may even have been ὁ ἐπὶ τῇ διοικήσει, before this magistracy became more prominent at the end of the 4th century B.C.[188] To my knowledge, Agyrrhios' law provides the earliest example of the διοίκησις in action in an Attic inscription.[189]

[187] Conomis (1970) has collected the ten surviving fragments of this speech, pp. 98–100, F 18–27.

[188] See Meritt 1960, pp. 3–4; Rhodes 1972, pp. 106–110; 1981, pp. 514–516.

[189] Prior to the discovery of this law, the earliest example seems to have been the inscription of the Lykourgan period cited in note 188, published by Meritt.

The syntax of the instructions for future action by the *apodektai* in the final three lines appears to be straightforward: [κ]αὶ τὸ λοιπὸν μὴ ⟨ἀ⟩φαιρεῖν τὼ δύο δεκάτ[.] ἐκ τῶν κατ⟨α⟩βαλλομένων χρημάτων, "for the future do not take away (subtract) the two *tenths* from the money that is being paid in." It is not immediately clear whether τοὺς ἀποδέκτας is to be understood as the subject of μὴ ⟨ἀ⟩φαιρεῖν or whether the instruction is a general one directed at whatever magistrates might be in a position to make such a subtraction. The former possibility seems to me more likely.

Again we do not know why anyone would want to subtract the two *tenths* from the money that is being paid in, nor is the identity of the latter transparent. Does the phrase ἐκ τῶν κατ⟨α⟩βαλλομένων χρημάτων[190] refer exclusively to payments that belong to the *dodekate* and the *pentekoste*, or does it include other kinds of payments as well? The maneuver of taking from the payments of these two taxes for the present year a sum equivalent to that produced by the two *tenths* last year and assigning it to the *dioikesis*—perhaps a piece of creative bookkeeping—is not to be repeated.

Throughout the final four lines of the law, I remain puzzled by the identity of the two *tenths* and their precise relationship to the *dodekate* on the *sitos* from the islands on the one hand and to the *pentekoste* on the other. Problems of interpretation of ἐκ τοῖν δυοῖν δεκάτ[.]|ιν in lines 58–59 and τὼ δύο δεκάτ[.] in line 60 begin with the fact that it is impossible to determine from the letters on the stone whether these nominal adjectives are feminine, masculine, or neuter. See Threatte 1996, pages 91–95, 448.

If they are feminine, denoting two *dekatai*, the reference could be to a pair of taxes or tithes. As we have seen above, page 27, notes 44, 45, the earliest known *dekate* tax at Athens appears to have been that imposed on the agricultural produce of Attica by Peisistratos. Unrelated to this is the next δεκάτη in our record: that found in the Kallias Decree, *IG* I³ 52A, line 7, καὶ τὰ (χρήματα) ἐκ τῆς δεκάτης ἐπειδὰν πραθεῖ ("and the money produced by the *dekate* whenever it has been sold"), 431 B.C.(?). This reference is too compressed and the context too controversial to permit certain identification beyond the facts that (1) the tax was computed in cash, (2) the right to collect it was "sold," and (3) the proceeds of that sale went to the gods. If δεκάτη here means exactly 10%, and not merely a "tithe" collected at a lesser rate, then this tax seems excessively onerous for the Athenians in the heyday of their empire to be imposing on their own citizens. Foreigners would be more likely targets.[191] The choice of the term δεκάτη may not be unrelated to the vote of the Athenian assembly prior to the Kallias Decree to earmark the revenue of this tax ἐς ἀπόδοσιν . . . τοῖς θεοῖς ("as repayment . . . for the gods"), line 5.

Mattingly and Cawkwell[192] have identified this δεκάτη as the forerunner of that levy collected in 410 B.C. by the Athenians under Alkibiades in Xenophon, *HG* 1.1.22, ἐντεῦθεν δ' ἀφικόμενοι τῆς Καλχηδονίας εἰς Χρυσόπολιν ἐτείχισαν αὐτήν, καὶ δεκατευτήριον κατεσκεύασαν ἐν αὐτῇ, καὶ τὴν δεκάτην ἐξέλεγον τῶν ἐκ τοῦ Πόντου πλοίων καὶ φυλακὴν ἐγκαταλιπόντες ναῦς τριάκοντα καὶ στρατηγὼ δύο, Θηραμένην καὶ Εὔμαχον, τοῦ τε χωρίου ἐπιμελεῖσθαι καὶ τῶν ἐκπλεόντων πλοίων καὶ εἴ τι ἄλλο δύναιντο βλάπτειν τοὺς πολεμίους ("From there they arrived at Chrysopolis in the territory of Kalchedon and fortified it and they constructed a tithe-collecting station there and they began to collect the *dekate* on ships sailing out of the Pontos, and leaving behind thirty

[190] Cf. *IG* II² 29; Tod, *GHI* II.116, lines 18–21, μερίσαι δὲ τ|ὸ ἀργύριον τὸ εἰρημένον τὸς ἀποδέ|κτας ἐκ τῶν καταβαλλομένων χρημά||[τ]ων ("The Receivers are to allocate the stated sum of money from the funds that are being paid in"), ca. 386 B.C.; Aristotle, *Ath* 48.1, ἀποδέκται δέκα . . . οὗτοι δὲ παραλαβόντες τὰ γ[ρ]αμματεῖα ἀπαλείφουσι τὰ καταβαλλόμενα χρήματα ἐναντίον τ[ῆς βου]λῆς ("There are ten Receivers . . . these officials after receiving the inscribed tablets erase the funds that are being paid in in the presence of the Council").

[191] This suggestion would become more attractive if we could follow the editors of *ATL* III, pp. 326–327, 334, endorsed by Mattingly (1964, p. 45), in concluding that the *Hellenotamiai* controlled the income from this tax, but Gomme (1953, p. 18, note 1) showed long ago that in reaching this view they have misconstrued the Greek in lines 4–7 of *IG* I³ 52A.

[192] Mattingly 1964; Cawkwell 1975, p. 54, note 4.

ships as a garrison with the two generals, Theramenes and Eumachos, to oversee the region and the ships that were sailing out and to inflict whatever other harm they could on the enemy"). This arrangement was revived by Thrasyboulos in 390/89 B.C. (*HG* 4.8.27), πλεύσας εἰς Βυζάντιον ἀπέδοτο τὴν δεκάτην τῶν ἐκ τοῦ Πόντου πλεόντων ("after sailing to Byzantion, he sold the *dekate* on goods sailing out of the Pontos"), 390/89 B.C., cf. *HG* 4.8.31; Diodoros 13.64.2.

In a famous passage, Polybios (4.44.3–4) refers to this same impost, ἐκ Καλχηδόνος γὰρ οἱ βουλόμενοι διαίρειν εἰς Βυζάντιον οὐ δύνανται πλεῖν κατ' εὐθεῖαν διὰ τὸν μεταξὺ ῥοῦν, ἀλλὰ παράγουσιν ἐπί τε τὴν Βοῦν καὶ τὴν καλουμένην Χρυσόπολιν, ἣν Ἀθηναῖοι τότε κατασχόντες Ἀλκιβιάδου γνώμῃ παραγωγιάζειν ἐπεβάλοντο πρῶτον τοὺς εἰς Πόντον πλέοντας ("Those who want to cross over from Kalchedon to Byzantion cannot sail on a straight course because of the intervening current but they make their way on an angle toward the Ox and what is called Chrysopolis, which the Athenians once seized on the advice of Alkibiades when they first set about collecting tolls on those sailing into the Pontos").[193]

Although Xenophon clearly speaks of ships coming out of the Pontos and Polybios of those sailing in, Boeckh (1886, I, pages 396–397) pointed out long ago that there is no contradiction between these two sources: "ohne Zweifel haben beide Recht, indem von beiden Ladungen, sowohl der in den Pontos, als der aus dem Pontos geführten, der zehnte erlegt wurde." For later explicit testimony regarding the *Zollstation* at Kalchedon collecting tolls on ships entering and leaving the Pontos, see the tax law of the province of Asia, ca. A.D. 62, Engelmann and Knibbe 1989, pages 46–48, lines 13–15.

Other ancient references to a *dekate* tax at Athens[194] are too general to be helpful in identifying the two *dekatai* in our lines 58–60. If the use of the dual, however, does in fact mark a close connection between these two taxes and if, as seems to be the case, they were familiar enough to the Athenians not to need any further identification than "the two *dekatai*" or "the twin *dekatai*," could they possibly be the double *dekatai*, so to speak, on goods moving into and out of the Pontos?

While this theory may have some possible verbal attraction, historical and chronological difficulties present themselves. We cannot assume that Athenian control of tolls and shipping at the entrance to the Bosporos survived the victories of Antalkidas and the Spartans in that region in 387 B.C. According to Xenophon (*HG* 5.1.28), ὁ δὲ Ἀνταλκίδας . . . ἐκράτει τῆς θαλάττης· ὥστε καὶ τὰς ἐκ τοῦ Πόντου ναῦς Ἀθήναζε μὲν ἐκώλυε καταπλεῖν, εἰς δὲ τοὺς ἑαυτῶν συμμάχους κατῆγεν ("Antalkidas was in control of the sea, with the result that he was able to prevent ships coming out of the Pontos to reach port in Athens and he diverted them to the allies of his own people").[195] It seems unlikely that the terms of the King's Peace would have allowed these Athenian imposts to remain in place. We have no evidence for successful restoration by 375/4 B.C. (πέρυσιν, line 58) of these "two *dekatai*" to the extent that revenue from them would have come into the Athenian public treasury.[196] On the other hand, Demosthenes 23.177 indicates that Athens was still deeply concerned about these tithes in 357/6 B.C. when Chairedemos of Oreus claimed them, ἐνθυμεῖσθ' ὅτι καὶ τέλη καὶ δεκάτας ἠξίου λαμβάνειν, καὶ πάλιν ὡς αὑτοῦ τῆς χώρας οὔσης τοὺς λόγους ἐποιεῖτο, τοὺς δεκατηλόγους ἀξιῶν τοὺς αὑτοῦ τῶν τελῶν κυρίους εἶναι ("Bear in mind that he justified collecting the taxes and the *dekatai* and that again he kept talking as if

[193] In his excellent note on this passage, Walbank (*HCP* I, p. 497) observes that "the phrase εἰς Πόντον, used here, and in iii. 2. 5 and in 52. 5 (based ultimately on a documentary source), suggests not that P. is merely writing carelessly, but that the toll was exacted on goods travelling in either direction."

[194] For these see *PCG* III.2, Aristophanes F 472; *IG* II^2 1609, II, line 97; Pollux 8.132; 9.28–31; Harpokration, *s.vv.* δεκατεύειν, δεκατευτάς, δεκατηλόγος. See above p. 27, note 45. Mattingly (1964, pp. 45–46) argues that Antiphon's use of the word δεκατευτάς (Harpokration) proves that the tithe at Chrysopolis was earlier than 410 B.C.

[195] For trouble at Byzantion at this time, which might suggest hostility to Athens, see Seager, *CAH*^{2} VI, p. 163.

[196] On Athenian finances in this period, see the comments of Wilson (1970), whose estimate of Athenian income from taxes, however, is restricted to the *eisphora*.

the territory belonged to him and he thought it right that the tithe-collectors should be his own people and in control of the taxes"). Note the plural δεκάτας.

If the two *dekatai* of Agyrrhios' law are to be sought elsewhere, a possible venue would have been in the three islands themselves. This was certainly the origin of the προκαταβολή in lines 55–56 and probably the πεντηκοστή in line 57. In defining a portion of these taxes equivalent to that fetched last year by the two *dekatai*, it might have been reasonable for the lawgiver to use two local taxes as his comparanda. Since our law concerns three islands, I do not understand why there are only two *dekatai* in lines 58–60. The existence of a pair of such taxes on Lemnos, however, could provide a clue to the puzzling fragmentary dual in line 10 of the Athenian decree of 387/6 B.C. governing land tenure and other matters in this klerouchy, *IG* II^2 30 (= *Agora* XIX, L3), κατὰ τὼ δ[ύο δεκάτω].[197] The context provides no information as to their identity (Myrina and Hephaistia?). But Demosthenes' general references in 24.120 to τὰ ἱερά, τὰς δεκάτας τῆς θεοῦ καὶ τὰς πεντηκοστὰς τῶν ἄλλων θεῶν ("sacred funds, *dekatai* of the goddess, and the 2% taxes of the other gods") and in 130 to τῶν δεκατῶν τῶν τῆς θεοῦ ("the *dekatai* of the goddess") should warn us against making hasty identifications. We have already seen that our fragmentary sources provide a completely inadequate picture of the complexity of the tax structure of Athens and her dependencies.

The possibility that the two duals in lines 58 and 60 are neuter or masculine could lead to more satisfactory solutions. As neuters, the two *tenths* might perhaps represent fractions of the *prokatabole* from the *dodekate* and the *pentekoste*. It has also been suggested to me that they might refer to land divisions on the islands, in other words, that Lemnos, Imbros, and Skyros were divided up into ten parts for purposes of taxation. I leave this difficult passage at the end of the law fully confident not only that my commentary on lines 55–61 fails to unravel all of their obscurities but also that very soon others will try to reach more plausible conclusions about these *tenths* and their financial context.

* * *

The text of the law ends abruptly in line 61. Although there was room on the stele, the cutter added no instructions for publication of the law. We know nothing as to how this was effected, who paid for the stele, and most importantly where it was originally erected. In this respect the surviving text of Agyrrhios' law differs markedly from that of Nikophon's law on silver coinage of the previous year, which devotes several lines to spelling out these details, *SEG* XXVI 72, lines 43–49, 55–56. It is unlikely that the full text in the archives containing the law on the grain-tax, which was promulgated by the *nomothetai*, completely lacked this vital information. Of the other eight surviving Athenian laws on stone from the 4th century B.C. only four have the end of their texts preserved; each one contains a publication formula (*IG* II^2 133; 140; *SEG* XII 87; XXVI 72).

One possible explanation for the omission of the publication formula on the grain-tax stele is that it was set up next to a number of other stelai containing enactments of the *nomothetai*. If several such *nomoi* were physically grouped together, repetition of the instructions for publication might have been judged unnecessary on each stele. The findspot of Agyrrhios' law in the northwest corner of the agora near the Royal Stoa, where many other stelai bearing Athenian laws were erected, may be suggestive.[198]

197 The isolated "word" [– – –]δεκατ[– – –] appears in the much later Athenian decree concerning Lemnos, *IG* II^2 1051, Fr. c, line 1, in a context that does not necessarily call for a date or a dedication. For a new edition of this inscription and its associated fragments, see Kallet-Marx and Stroud 1997. For a possible double δεκάτη, one on dry produce, the other on wine, see *SEG* XXXVIII 380, from Chaironeia, with the discussions of Feyel 1942 and Knoepfler 1988, pp. 290–294.

198 For laws published on stelai and on walls at the Royal Stoa, see *SEG* XL 146, with earlier bibliography. Kevin Clinton has suggested to me that the omission of the publication formula, which is in keeping with the absence of many other important details in the text of our law, could be another indication that this stele is just one of a series of laws on the *dodekate*, all subordinate to a master law on this tax. The latter may have spelled out the details missing from the law on Lemnos, Imbros, and Skyros, and contained a publication formula covering all the stelai in the series. I hope that he will develop this and his other ideas on the grain-tax law in print.

CHAPTER III

THE AIAKEION

THIS CHAPTER takes up three aspects of the physical setting of the grain-tax law, which concern (1) the history, identification, and location of the temenos of Aiakos, (2) the transportation of the people's wheat and barley from the Peiraieus to the Aiakeion, and (3) the venue in which the public sale of the grain took place.

HISTORY, IDENTIFICATION, AND LOCATION

The law provides important new information on the Sanctuary of Aiakos which now permits a radically different identification of a large building in the Athenian Agora. Our earliest testimony for a sanctuary of this Aiginetan hero in Athens comes from Herodotus.[1] In an extended account of the military challenges confronting the new Kleisthenic state, 5.65–96, he tells that the Thebans, when hard-pressed by the hostility of the Athenians, responded to a Delphic oracle by appealing to the Aiginetans. The latter sent them images of the Aiakidai as helpers. When these proved ineffectual in battle and the Athenians were victorious, the Thebans sent the images back and asked the Aiginetans for men instead (5.79–81). The Aiginetans responded this time by attacking the coast of Attica, which prompted the beleaguered Athenians themselves to consult Delphi. Apollo told them, according to Herodotus, to desist from hostilities against the Aiginetans for thirty years, and then in the thirty-first year to lay out a temenos of Aiakos in Athens and begin the war against the Aiginetans only at that point: ἐπισχόντας ἀπὸ τοῦ Αἰγινητέων ἀδικίου τριήκοντα ἔτεα, τῷ ἑνὶ καὶ τριηκοστῷ Αἰακῷ τέμενος ἀποδέξαντας ἄρχεσθαι τοῦ πρὸς Αἰγινήτας πολέμου ("holding off from [avenging] the injustice of the Aiginetans for thirty years, in the thirty-first year, after laying out a temenos to Aiakos, begin the war against the Aiginetans") (5.89.2). If they followed this advice, they would succeed. If, however, they attacked at once, they would eventually prevail but only after great suffering.

The Athenians chose not to follow the oracle's bidding. They immediately laid out a temenos of the Aiginetan hero and began preparations to wage war against the Aiginetans (5.89–90). Their war plans were interrupted, however, by the threat of a Peloponnesian invasion mounted by the Spartan king Kleomenes for the purpose of restoring the tyrant Hippias in Athens. This threat never materialized, and Herodotus does not return to the narrative of the struggle between Athens and Aigina until 6.49–50, 85–94, in 491 B.C.

For our purposes it is important to note that the historian is explicit not only on the fact that the Athenians laid out Aiakos' shrine without delay, but also that the temenos was at or near the Agora and still visible in Herodotus' day: ταῦτα ὡς ἀπενειχθέντα ἤκουσαν οἱ Ἀθηναῖοι, τῷ μὲν Αἰακῷ τέμενος ἀπέδεξαν τοῦτο τὸ νῦν ἐπὶ τῆς ἀγορῆς ἵδρυται ("When the Athenians heard these

[1] For Aiakos on Aigina, see Welter 1938, pp. 119–120; 1962, pp. 88–90; Hubbard 1987; Zunker 1988, pp. 45–46, 63–64, 69–72; Rutherford 1992; Carnes 1995. Aiakos' kinship to Herakles is briefly mentioned in an uncertain context in the fragmentary inscription *IG* II^2 885, line 17, which Allen (1971) argued came from Aigina, not Athens. He also urged that the Pergamene king Attalos I was σύνναος with Aiakos in the Aiakeion on Aigina. See also Figueira 1993, pp. 396–397. Presumably all physical trace of such an association had disappeared before Pausanias described the Aiginetan Aiakeion; see p. 92.

For Aiakos in general, see Apollodoros, *Library* 3.12.6, with J. G. Frazer's copious notes in the Loeb edition *ad loc.*; Zunker 1988, pp. 63–89. J. Boardman, *LIMC* I.1, pp. 311–312, *s.v.* Aiakos, has collected representations in ancient art, to which should be added the sherd mentioned below in note 21.

instructions, they laid out a temenos for Aiakos, the present one that is established on the Agora") (5.89.3). With the exception of the troublesome 5.67, which I hope to discuss elsewhere, all of the other twenty-nine examples of the word τέμενος in Herodotus designate an open precinct that may or may not contain buildings.

It is clear from the sequence of Herodotus' narrative that he regarded these events as having taken place soon after the reforms of Kleisthenes but before the Ionian revolt and the arrival of Aristagoras in Athens, that is, between ca. 507 and 499 B.C. He frames his narrative of the founding of the temenos of Aiakos, in a familiar manner, with two passages that refer to the visit of Aristagoras. In 5.65.5 Herodotus says: οὕτω μὲν Ἀθηναῖοι τυράννων ἀπαλλάχθησαν. ὅσα δὲ ἐλευθερωθέντες ἔρξαν ἢ ἔπαθον ἀξιόχρεα ἀπηγήσιος, πρὶν ἢ Ἰωνίην τε ἀποστῆναι ἀπὸ Δαρείου καὶ Ἀρισταγόρεα τὸν Μιλήσιον ἀπικόμενον ἐς Ἀθήνας χρηίσαι σφέων βοηθέειν, ταῦτα πρῶτα φράσω ("In this way the Athenians rid themselves of their tyrants. But all the things they did and suffered that deserve description after they were freed, before the revolt of Ionia from Dareios and before the arrival of Aristagoras the Milesian in Athens to ask for their help, these I will first describe").[2]

At the end of this section, he returns to Aristagoras' embassy to Athens: ἐν τούτῳ δὴ τῷ καιρῷ ὁ Μιλήσιος Ἀρισταγόρης, ὑπὸ Κλεομένεος τοῦ Λακεδαιμονίου ἐξελασθεὶς ἐκ τῆς Σπάρτης, ἀπίκετο ἐς τὰς Ἀθήνας ("At this very time the Milesian Aristagoras, sent away from Sparta by the Lakedaimonian Kleomenes, arrived in Athens") (5.97.1). Within his narrative, as defined temporally by these two framing passages, Herodotus includes several flashbacks, as is his wont. In only three places, however, does he refer forward to later events, that is, to events later than the Ionian revolt. In each case he does so with regard to customs or monuments that originated before the revolt but survive into his own day: (1) Argive and Aiginetan women still wear longer brooch-pins in their clothing because of the ancient feud with Athens (5.88.3); (2) the fetters with which the Athenians bound the Boiotian prisoners in 506 B.C. still hang on a fire-charred wall on the Acropolis where Herodotus later saw them (5.77.3); (3) the temenos of Aiakos that they laid out at the time of their war with Aigina is the one that is still there near the Athenian Agora (5.89.3). Throughout this whole section in 5.65–96, Herodotus never places any other event in his narrative later than the Ionian revolt.

I have rehearsed these features of Herodotus' narrative at some length for two reasons. First, the historian provides a valuable piece of chronological information to aid us in our search for the Aiakeion. Any archaeological candidate must have been built before Xerxes' invasion and destruction of Athens and have been standing on or near the Agora in Herodotus' day. Second, despite the fact that the historian explicitly states that the Athenians did not wait thirty years (τριήκοντα δὲ ἔτεα οὐκ ἀνέσχοντο, 5.89.3), the hypercritical wing of Herodotean scholarship has, predictably, rejected his chronology, rationalized the oracle, and rewritten the history of these events to conform to its own views. Many scholars have even claimed that since Herodotus' Aiakeion is unlikely to have survived the Persian invasion, it must have been built later.[3] This, like

[2] For the structure of Herodotus' narrative and chronology of this war with Aigina, see Andrewes 1937; Hammond 1955, pp. 406–411; Immerwahr 1966, pp. 116–119; Figueira 1981, pp. 5–7; 1991, p. 104.

[3] Typical representatives of this approach to ancient history are Macan (1895, p. 233): "But is the oracle correctly dated? The cult of Aiakos is suggestive of Philaid auspices: the actual date of the conquest of Aigina is 456 B.C. ... Thirty-one years from that date carries us back to 486–7 B.C. [*sic*] the probable date of the great Aiginetan war, misplaced by Hdt. ... It does not seem likely that the shrine of Aiakos in the Agora, seen by Hdt., was a survival from the pre-Persian days ... The oracle ... belongs presumably to a date just about the time when Athens was at war with Aigina ... when Kimon perhaps was in exile, and his recall being agitated. A descendant of Aiakos was the proper man to bring the Aiginetans to reason." Cf. How and Wells 1928, II, pp. 49–50: "H. is probably guilty of an anachronism in dating the project and the oracle before the Ionic Revolt ... The shrine of Aeacus in the Agora could hardly have escaped the ravages of the Persians in 480 B.C. May it not be (like the oracle) of later date, connected perhaps with Cimon ... who as a Philaid traced descent from Aeacus?" This approach has a distinguished pedigree in Köhler 1891; Wilamowitz 1893, II, pp. 280–281; Busolt 1895, II, pp. 644–645, note 3;

many such attacks on Herodotus, is special pleading. Shear (1993, pages 418–429) has cogently and persuasively demonstrated, on the basis of stratigraphy and context pottery, that some pre-Persian monuments in the Agora like the Old Bouleuterion, Building F under the Tholos, and possibly the Stoa Basileios were rebuilt after suffering damage at the hands of the Persians. There is no good reason why the Aiakeion could not have been another. At the risk of incurring the familiar charge of naiveté or fundamentalism, I prefer to conclude that on this point at any rate Herodotus knew what he was talking about. I believe that he spent considerable time in Athens and that his method here may have resembled that so well formulated by Jacoby in a slightly different context: "His is the manner of the Ionian ἱστορίη which, in front of the monuments, asks for their history and receives it from the λόγιοι ἄνδρες."[4] We must remember that in Herodotus' oral and earliest reading audience were many Athenians who knew when and why they and/or their fathers had built the temenos of Aiakos. For them, the historian's demonstrative pronoun, τέμενος ἀπέδεξαν τοῦτο τὸ νῦν ἐπὶ τῆς ἀγορῆς ἵδρυται, may also have had a special point, although it has gone largely unremarked by editors and students of Athenian topography.[5]

Before we leave Herodotus, it may be useful to glance at the only other two passages where he mentions Aiakos. In 6.35 he observes that the Philaid dynast Miltiades, son of Kypselos, a contemporary of Peisistratos, traced his ancestry back to Aiakos and Aigina. The line descended through Telamon, son of Aiakos, to Ajax, son of Telamon, to Philaios, son of Ajax and brother of Eurysakes. According to Plutarch, Solon pressed Athens' claim to Salamis against Megara partly on the grounds that when Philaios and Eurysakes moved from the island to Attica, they handed Salamis over to Athens. Philaios, eponym of the family and of the later deme of the Philaidai, settled in Brauron, while Eurysakes dwelt in Melite. Philaid connections with Aiakos, therefore, can be traced back in Athenian tradition at least to the early 6th century B.C.[6] It is noteworthy that no surviving source says anything in any of these traditions about Aiakos himself moving to Athens at any time, either alone or with his great-grandsons Eurysakes and Philaios, before ca. 506 B.C.

In 8.64 Herodotus relates that on the day before the battle of Salamis, after an earthquake, the Greeks on the island prayed to all the gods and called for aid from the local heroes, Ajax and Telamon. They also sent a ship to Aigina to fetch "Aiakos and the other Aiakidai." Macan (1895) noted *ad loc.*, "In view of the story in 5.80, 81 it may fairly be argued that this ship was sent to fetch

---

Beloch 1914, II, p. 25, note 3. For more of the same, see Walker, *CAH* IV, p. 258, unfortunately followed by Wycherley in *Agora* III, p. 48, who, without discussion, speaks of Herodotus' "incomplete and confused account of the relations of Athens and Aegina" and dates the dedication of the Aiakeion at Athens ca. 488 B.C., i.e., "thirty years before the Athenians actually inflicted a decisive defeat on Aegina." Similarly, Nenci 1994, pp. 283–284. The prize in this ongoing derby of outsmarting Herodotus, however, should so far go to Oikonomides (1990, p. 22), who manages to place the oracle and the building of the Aiakeion "*ca.* 488," thereby permitting the latter to stand only until the Persian invasion when "from the excavations we know that they [see below, p. 102] were then destroyed by fire." This is to make the *pater historiae* into both a liar and a fool. It is instructive to see modern historians taking greater liberties with the Delphic oracle than did the ancients. One is reminded of A. W. Gomme's wry observation in another context: "This is very characteristic of the way 'we modern historians' correct the ancients ... We begin by rejecting the authority of Herodotus and Thucydides, and then, having got rid of what is almost our only evidence, pretend that we know all about the political conditions in 479" (1945, I, pp. 268–269). Pritchett (1993) has assembled a chamber of horrors of this kind of scholarship. In my view there is nothing to recommend the notion of Podlecki (1977) that Herodotus did not visit Athens.

[4] Hellanikos, *FGrH* III B, 323a F 1, Supplement vol. I, p. 22.

[5] *Pace* Nenci as cited in note 3, this demonstrative pronoun has not been "molto discusso"; it deserves more attention than I can give it here. I do not find persuasive the briefly stated view of Schwartz that this passage was written later in Thourioi since "un lecteur athénien n'aurait sans doute pas eu besoin que l'on précise par τὸ νῦν ἐπὶ τῆς ἀγορῆς ἵδρυται" (1969, p. 369).

[6] Plutarch, *Solon* 10.3. Pausanias briefly notes this family connection in 1.35.2, where he makes Philaios son of Eurysakes, and 2.29.2–4. See also *Vita Marcellini Thucydidis* 2–4. For discussions of how and why the Philaidai were originally connected with Salamis and Aiakos, see Ferguson 1938, pp. 15–20; Prinz 1979, pp. 34–56; Thomas 1989, pp. 161–173; Taylor 1997, pp. 21–47.

actual idols (ξόανα), and not merely to perform an invocation *in loco*." In 8.83–84 the ship, here called a trireme, brought the Aiakidai over to Salamis, and the Aiginetans later claimed that this was the ship that began the naval battle.[7] Our sources do not identify the place in Aigina whence the statues of the Aiakidai were removed on this occasion, but their most plausible permanent home was the Aiakeion in the city, on which we shall have more to say presently.[8] Presumably, the images went back to Aigina after the battle since Herodotus says no more about them. His silence here and in 5.89 makes it unlikely that the images remained on Salamis as some kind of memorial to their timely aid in the Greek victory, or that the returning Athenians in 479 B.C. were somehow able to bring them to the Agora and set them up in the Aiakeion.

These passages provide a little background to the founding *aition* of the Aiakeion on the Athenian Agora by showing that the Aiginetan hero had ties with at least one prominent old Athenian family, the Philaidai, and that the well-traveled images of the Aiakidai had on one occasion, at any rate, made their way to Attic soil.

Although neither Herodotus nor any other source mentions images, bones, a grave, an altar, or indeed anything in the Aiakeion at Athens, Kearns is undoubtedly right in interpreting the Athenians' motive in laying out the temenos ca. 500 B.C. as an attempt in wartime to summon away the most powerful hero of an enemy state: "The hero must in some sense be brought to Athens."[9] In this case we know something about his parent shrine on his native island from Pausanias' description, which I examine below, pages 92–93. According to the periegete's informants, Aiakos' bones were indeed buried there, hidden below his altar. We hear of no attempt on the part of the Athenians to seize them at this or any other time, as Kimon did later with the bones of Theseus. I argue below that the Athenians tried to make Aiakos feel at home in Athens by constructing a temenos that closely resembled his Aiginetan abode. Finally, Herodotus' story, as persuasively interpreted by Kearns, implies that before ca. 506 B.C. Aiakos was "living" in Aigina and had not yet taken up residence in Athens.

The next reference to the Aiakeion in Athens probably comes in Stele VI of the "Attic Stelai," *IG* I^3 426, lines 5–8, which records the confiscation and sale of a house once belonging to one of the mutilators of the Herms, 414 B.C.

[. . .ca. 7. . .]ο τõ Διοδόρο Ει̣[τεαίο]
[οἰκ]ία ἐν Κολλυτõι *h*ε̃[ι γεĩτον]
ἐκ τõ ἐπὶ θάτερα τὸ Αἰ[– – –]
καὶ *h*ε ἀγορά

A house of Diodoros of Eiteia in Kollytos
which is bounded on one side by the
Ai[– – –] and on the other by the Agora.

[7] In the epiphany reported by some at the time of the battle, the Aiakidai resembled armed men coming from Aigina who stretched out their hands to protect the Greek triremes (Plutarch, *Themistokles* 15.2).

[8] Below, pp. 92–93, 101–102. For the Aiakidai, see Zunker 1988, pp. 71–72; Rutherford (1992) has argued persuasively that Pindar Fr. Pa. XV (= *POxy* 2441) was written for an Aiginetan chorus to perform at an annual procession in which statues of Aiakos and Nereus were carried on a wagon from the Aiakeion on Aigina to the sanctuary of Zeus Hellanios on Mount Hellanios. See also Zunker 1988, pp. 72–78; Carnes 1995, pp. 17–21. I have been unable as yet to consult J. S. Carnes, *The Uses of Aiakos: Pindar and the Aiginetan Imagination*, Lanham, Md., forthcoming.

[9] Kearns 1989, p. 47; cf. independently, Figueira 1991, p. 104: "Just as the Eurysakeion, a hero shrine of the Athenian *genos* of the Salaminioi, solidified an Athenian claim to the ownership of the island of Salamis, the Aiakeion expressed a similar claim to Aigina. If the descendants of Aias could make Salamis over to Athens, why could they not transfer a title to Aigina, inherited from Telamon, son of Aiakos?" See also Figueira 1993, pp. 93, 211, 277, 279, 296–297, 404, for similar sentiments. Parker 1996, p. 157, note 18; Williams 1987, pp. 672–674, "Athens' decision to set up a sanctuary of Aiakos in Athens is a clear attempt to subvert the power of Aiakos and represents the beginning of a propaganda war with Aigina" (p. 672).

In his editio princeps of this fragment, which was found in the Agora Excavations, Pritchett restored the name of the sanctuary in line 7 as τὸ Αἰ[άκειον]. Since the house in the deme of Kollytos is said to have bordered on the Agora, Pritchett combined this new topographic information with other sources to argue that Kollytos lay "in part north of the gap between the Pnyx and the Areopagos," that is, near the southwest corner of the Agora. Accordingly, he concluded that the Aiakeion also "must now be sought at the southwest corner of the Agora."[10]

Lewis (1955, page 16) rejected this restoration in line 7 in favor of τὸ Αἰ[άντειον] on the grounds that "the cult-centre for Aias was the Eurysakeion, and Wycherley will show that there is no difficulty in identifying the Eurysakeion with the Aianteion. If this is so, the Aianteion will fit the topographical requirements of the passage, and we will still be without firm evidence for the site of the Aiakeion."

Wycherley printed Pritchett's restored text of *IG* I^3 426, lines 5–8, among his testimonia for the Aiakeion in *Agora* III, pages 48–49, but, without reference to Lewis' publication, he noted a suggestion of Eliot that Αἰ[άντειον] was a possible reading, "if indeed there was an Aianteion at Athens." As Lewis predicted, Wycherley also attempted to establish a connection between Ajax and the shrine of his son, the Eurysakeion, on the edge of the Agora in Melite, but the task of demonstrating that the latter was also called the Aianteion proved to be more formidable (*Agora* III, pages 90–93, 225), although he succeeded in convincing Kron (1976, pages 173–174) and Kearns (1989, pages 82, 141).

To my knowledge, the word Αἰάντειον is attested in Athens only once. It is found in a decree of the demos of the Salaminians of 118/7 B.C. praising and crowning the Athenian ephebes of 119/8 B.C. for their participation in the parade, sacrifice, and games of the Aianteia festival on Salamis. The crown is to be proclaimed at the tragic contests of the Dionysia on Salamis. The decree is to be inscribed on a stone stele and set up ἐν τῷ Αἰαντείῳ (*IG* II^2 1008, lines 75–88).

[10] Pritchett 1953, pp. 275–276, who was thus able to bring this critical new piece of evidence to bear on the still valuable discussion of Kollytos in Judeich 1931, p. 169 and on Young's (1951, pp. 140–143) somewhat inconclusive attempt to decide whether the "industrial district" that he excavated between the Hill of the Nymphs and the Areiopagos belonged to Melite (which he favored) or Kollytos. Lewis (1955, pp. 16–17) usefully remarks on the contiguity of these two demes, based on Strabo 1.65. Judeich well observed that Plutarch's tag, οὐδὲ γὰρ Ἀθηναῖοι πάντες κατοικοῦσι Κολλυτόν ("For certainly all Athenians do not live in Kollytos"), *De exilio* 6; *Moralia* 601B, is vividly illustrated by the luxury homes excavated on the slopes of the Areiopagos and the Hill of the Nymphs looking down into the narrow valley below; cf. also Young 1951, p. 139.

A location for Kollytos stretching to the south from the Agora between these two hills would fit the observation of Himerios preserved in Photios, *Bibliotheca*, p. 375B (Bekker), στενωπός τις ἦν Κολυτός, οὕτω καλούμενος ἐν τῷ μεσαιτάτῳ τῆς πόλεως, δήμου μὲν ἔχων ἐπώνυμον, ἀγορᾶς δὲ χρείᾳ τιμώμενος ("there was a narrow passage called Kollytos in the most central part of the city which both gave its name to a deme and served the purpose of an agora"). Unaccountably, while citing the evidence of *IG* I^3 426, lines 5–8 for the location of Kollytos, Traill (1975, p. 40 and maps 1–3) placed this deme on the south side of the Acropolis, far from the Agora, in both his "Conspectus of Deme Locations" and on his maps. In Traill 1986, p. 126 and map, citing the same evidence, the author places Kollytos "SW of Agora, S of Areopagos" and is followed by Walbank (1994, p. 138). W. E. Thompson (1970, pp. 66 67), without evidence, suggested that Kollytos lay to the north of the Agora, but Judeich, *loc. cit.*, long ago showed that this is impossible. All of the houses excavated in the industrial district by Young and his colleagues are too far south to qualify for identification as the οἰκία in Kollytos mentioned in this passage of the Attic Stelai.

S. B. Aleshire has drawn my attention to the sanctuary of Eileithyeia, which a new fragment of *IG* II^2 1590 + 1591 (*Agora* XIX, L6; *SEG* XXXIII 167), col. II, lines 97–98, tells us lay in Kollytos. If the Eileithyeion was below the northeast lower slopes of the Acropolis, as Pausanias 1.18.4–6 implies and the majority of the finding-places of her dedications seem to suggest, then we should think of Kollytos as extending also along the lower northern slopes of the Acropolis south of the Agora. For helpful bibliography on the Eileithyeion, see Walbank 1983, p. 119. For the relationship of Kollytos to Diomeia, see Billot 1992, pp. 124–129. To my mind Wallace (1992, p. 329) shows that the argument linking the Olympieion and the house of Charmides in Andokides 1.16 with the location of Kollytos is not persuasive, *pace* his supposition on p. 333. I intend to return elsewhere to the question of the location and extent of Kollytos.

An almost identical decree of the demos of the Salaminians of 127/6 B.C., praising the ephebes' activities on Salamis, was also inscribed on a stele set up ⟨ἐ⟩ν τῶι τεμένει τοῦ Αἴαντος (Reinmuth 1955, page 231, lines 129–142).

In introducing his discussion of these two documents as evidence for an Aianteion in Athens, Wycherley observed, "These references may well be to the shrine at Salamis, in which case they are not relevant" (*Agora* III, page 91). This, I believe, is clearly so.[11] While Ajax may have been closely associated with his son in the Eurysakeion, we do not yet have any evidence that the Athenians ever called that sanctuary τὸ Αἰάντειον. Ajax' main sanctuary was on Salamis, where his festival, the Aianteia, was celebrated. It was there that the demos of the Salaminians erected their honorary stelai for the ephebes. Copies of their decrees formed parts of dossiers of several decrees honoring the ephebes which were inscribed on stone stelai and set up in the Athenian Agora in or near the Eurysakeion. But the latter setting and the Aianteion on Salamis were, as far as we know, two widely different places.[12]

Not only does the restoration Αἰ[άντειον] in the Attic Stelai lack a valid parallel, but it suffers from another disability, perhaps, not fatal, but at least troubling. The Eurysakeion, with which Lewis, Eliot, and Wycherley equate the alleged Aianteion, was in the deme of Melite, probably near Kolonos Agoraios and the Agora.[13] The sanctuary we seek to restore in the text of the Attic Stelai, however, stood next to a house in Kollytos. This may not be an insuperable problem, since we know that Kollytos and Melite were contiguous. But since the only reason for bringing Melite into the discussion in the first place arose from the erroneous restoration τὸ Αἰ[άντειον], it is better to leave aside Melite, the Eurysakeion, and the rest, and concentrate on the evidence we have.

What we have is a house in Kollytos which bordered the Agora on one side. On the other side was a sanctuary of a deity or hero whose name begins with Αἰ[– – –]. In Wycherley's collection of testimonia in *Agora* III there are more than forty different gods and heroes known to have been worshiped in the Agora. With the exception of Ajax, Aiakos is the only one whose name begins with Αἰ[– – –]. To this formal criterion for restoration we can add the certain knowledge that his sanctuary existed earlier in the 5th century B.C., when Herodotus was writing, and in 374/3 B.C., as the grain-tax law now attests. Until someone can propose a more plausible restoration, it seems to me that there are sound reasons for accepting Pritchett's τὸ Αἰ[άκειον] in this text of 414 B.C. We will return to the probable location of this house in Kollytos, but for the present, we should also bear in mind Pritchett's inference that the Aiakeion is to be sought near the southwest corner of the Agora.

Three lexicographical notes complete the dossier of epigraphic and literary testimonia for the Aiakeion. The earliest of these escaped the net of Wycherley in *Agora* III but not the sharp eye of Oikonomides (1990). To my knowledge no other student of Athenian topography has mentioned this evidence. Oikonomides aptly drew attention to a fragment of a lexikon of the 2nd century A.C. preserved on a mutilated papyrus from Oxyrhynchos (*POxy* 2087). In the editio princeps of 1927 Hunt pointed out that although the compiler's criteria for selection are unclear, his glossary of comparatively rare words seems to have been drawn from Classical prose authors

[11] This is also the view of Pélékidis (1962, pp. 248–249).

[12] In my view Pausanias 1.35.3 does not show that Ajax and Eurysakes were worshiped together in the Eurysakeion. While Wycherley (*Agora* III, pp. 90–91) rightly questioned the assertion of Ferguson (1938, p. 18), "On being adopted in 508/7 B.C. [as a tribal hero] he was accommodated in the temenos of his son Eurysakes," it is clear from Agora I 3625 (= *Agora* III, no. 255) that this shrine was an important center for the tribe Aiantis. Cf. Langdon 1987. The legends, testimonia, location, and possible archaeological remains of the Eurysakeion merit careful restudy—another excellent topic for a Ph.D. dissertation.

For Ajax and Eurysakes on a stele from Salamis, see *SEG* XL 130, with the important corrections of Taylor (1995). In a useful and perceptive exploration of Eurysakes and the future hero cult of his father, Henrichs (1993) reminds us that while Ajax had a temple and temenos on Salamis, his tomb was in the Troad (Strabo 13.1.30, C595; Pausanias 1.35.5).

[13] *Agora* III, pp. 90–93, nos. 246–255.

such as Herodotus, Thucydides, Plato, Demosthenes, and Aristotle.[14] The entry that concerns us is not attributed to a specific author. After examining the papyrus in the Ashmolean Museum in Oxford, I have devoted a separate paper to establishing a new text of the lines in question, which I believe should be read as follows:[15]

16 Αἰά[κ]ι̣ον κ(αὶ) ἡ Θόλο̣ς̣ ο̣[ὗ] φασι [τ]ὸν Αἰακὸν
17 οἰκῆσ̣[α]ι· Θό[λο]ς̣ δ̣(ὲ) ὅπου δει̣[π]ν̣ῖ ἡ̣ πρυτα-
18 νε(ύ)ου[σ]α φυλ̣ή̣. [ἐν δ(ὲ)] τ̣ῷ Αἰακ̣ί̣ῳ δίκ(αι) ἀ(να)γράφον/ται̣.

Aiakeion and the Tholos: where they say that Aiakos used to dwell; now the Tholos is where the prytanizing tribe dines; [whereas in/at] the Aiakeion *dikai* are published.

The supreme importance of this passage is that either the compiler of the lexikon or, more probably, his source closely linked the Aiakeion with the Tholos. The position of the latter in the southwest corner of the Agora, fixed by excavation, is therefore potentially important for anyone seeking to locate the Aiakeion, especially since Pritchett had already pointed us in that direction. The exact details of the bond between these two ancient structures remain obscure. That it may be a bond, however, closely tied to the location of the Aiakeion and the Tholos is suggested, at a minimum, by the two locative conjunctions in the papyrus, οὗ and ὅπου, which introduce parallel defining clauses. Briefly to summarize the suggestion I have made elsewhere at greater length, the most natural meaning of the formulation, "Aiakeion and the Tholos: where they say that Aiakos lived; now the Tholos is where . . .," is, in my view, that the two buildings were separate and were standing at roughly the same time. I believe that the compiler of the lexikon found Αἰάκειον καὶ ἡ Θόλος in his Classical (prose?) source and defined each with a locative clause. The Aiakeion was the dwelling of Aiakos where lawsuits were also written up; the Tholos was the place where the prytanizing tribe dined. The absence of the Tholos from Herodotus' brief discussion of the Aiakeion in 5.89, cited above, probably rules out the historian as our compiler's Classical source. The formulation "Aiakeion and Tholos" would not be unnatural in a Classical author, however, since we know that the Aiakeion, laid out ca. 506–499 B.C., was seen by Herodotus in the Agora in the second half of the 5th century B.C. and served as a public granary after 374/3 B.C. The Tholos, constructed ca. 460 B.C. or earlier, survived until late antiquity.[16] They may even have been coupled by a Classical author because they stood next to each other.

To my knowledge, information on the Aiakeion appears next in Greek literature in the *Lexikon* of Hesychios: Αἰάκειον· οὗ φασιν Αἰακὸν οἰκῆσαι (Latte 1953), A 1658. The interest of this item, which seems clearly to derive from the same tradition as *POxy* 2087, is that Hesychios and/or his source has not here included any information on the Tholos. He seems firmly to have divided Aiakeion from Tholos, treating them as two different structures. Likewise, in his entry on the Tholos, (Latte 1953), Θ 634–635, Hesychios makes no mention of Aiakos or the Aiakeion.

A few lines above the entry for Aiakeion in A 1658, the following lemma appears in Hesychios A 1653: AIAIAKON TION Ἀθήνησι . . . καὶ τὸ Αἰακοῦ τέμενος. Various emendations have been proposed; Schmidt and Latte have also both tried to combine these two separate lemmata

[14] Hunt 1927, pp. 110–113, no. 2087; Körte 1932, pp. 230–231, no. 770; Pack 1965, p. 115, no. 2120.

[15] Stroud 1994, where I printed a photograph and a detailed commentary on the readings. There I also argued that καί in line 1 is copulative and not epexegetical.

[16] On the date of the Tholos, see Thompson and Wycherley, *Agora* XIV, p. 42. In Stroud 1994 I tried to show that this text provides no support for the theory of Oikonomides (1990) that this passage demonstrated a close connection between the meals served in the Tholos and the hero Aiakos. I argue below, pp. 102–104, that there are also valid archaeological and historical grounds for similarly rejecting his belief that the structural remains excavated below the foundations of the Tholos belonged to "the house of Aiakos." In my view the perceptive analysis of Schmitt-Pantel (1992, pp. 168–177) of the origins of the communal meals of the post-Kleisthenic prytaneis makes very unlikely any possible connection between the Tholos and Aiakos.

into a single composite note, *viz.*, Αἰάκειον οὗ φασιν Αἰακὸν οἰκῆσαι Ἀθήνησι καὶ τὸ Αἰακοῦ τέμενος (Schmidt) and Αἰάκειον· τόπος Ἀθήνησι οὗ φασιν Αἰακὸν οἰκῆσαι καὶ τὸ Αἰακοῦ τέμενος (Latte). Neither the lemma in A 1653 nor these two composite texts add substantially to our information about the sanctuary of Aiakos.[17]

Finally, in Bekker, *Anecdota Graeca* 1.212.15, we find: Αἰάκιον· τόπος οὗ φασι τὸν Αἰακὸν οἰκῆσαι; cf. Photios, α-500 (ed. Theodorides); Bachmann, *Anecdota Graeca* 1.49.4. Herodian briefly notes the spelling and accentuation Αἰάκειον in Lentz 1867–68, III.1, page 375.12; III.2, pages 460.11, 848.3–4.

Prior to the discovery of the grain-tax law, these were all the literary and epigraphic testimonia regarding the Aiakeion in Athens. Before we assess our new information, it may be instructive to consider one piece of comparative evidence. While it is disappointing to find no mention of the temenos of Aiakos in Pausanias' description of Athens, he has left us an account of the hero's shrine on Aigina. From it we can at least gain an impression of the form and appearance of the parent sanctuary, which is the only other sanctuary of Aiakos mentioned in surviving ancient sources: ἐν ἐπιφανεστάτῳ δὲ τῆς πόλεως τὸ Αἰάκειον καλούμενον, περίβολος τετράγωνος λευκοῦ λίθου. ἐπειργασμένοι δέ εἰσι κατὰ τὴν ἔσοδον οἱ παρὰ Αἰακόν ποτε ὑπὸ τῶν Ἑλλήνων σταλέντες. . . . τοῦ περιβόλου δὲ ἐντὸς ἐλαῖαι πεφύκασιν ἐκ παλαιοῦ καὶ βωμός ἐστιν οὐ πολὺ ἀνέχων ἐκ τῆς γῆς. ὡς δὲ καὶ μνῆμα οὗτος ὁ βωμὸς εἴη Αἰακοῦ, λεγόμενόν ἐστιν ἐν ἀπορρήτῳ ("In a very prominent part of the city lies what is called the Aiakeion, consisting of a square peribolos of white stone. Worked in relief at the entrance are representations of those who were once sent to Aiakos by the Greeks . . . inside the peribolos olive trees have been growing for a long time and there is an altar that does not project much above ground level. There is a secret tradition that this altar is in fact the grave of Aiakos") (2.29.6–8).

It is interesting to note the prominent location of the Aiakeion on Aigina and that it consisted of an unroofed, square enclosure, built of white stone, and large enough to accommodate ancient—and therefore, presumably, fairly large—olive trees. It clearly had a prominent decorated entrance. Although it had a low altar, any earlier *naos, adyton,* or other structure inside the peribolos had presumably been removed by Pausanias' day. We have seen from Herodotus 5.89 that the Athenians too built their Aiakeion in a central position, ἐπὶ τῆς ἀγορῆς, and that it was probably an open precinct.[18]

[17] Wycherley (*Agora* III, no. 48), without having *POxy* 2087 in mind, was uncomfortable about trying to merge these two separate lemmata in Hesychios, probably for good reasons. In A 1658, the locative clause οὗ φασιν Αἰακὸν οἰκῆσαι clearly belongs to the lexicographical tradition of *POxy* 2087, where the words Ἀθήνησι, τόπος, and τέμενος do not appear. Conversely, there is nothing in A 1653 that resembles the locative clause of A 1658. It is probably better to keep the two notes separate and to consider the possibility that Hesychios or his source drew them from two different origins. As H. A. Thompson (1953, p. 45, note 28) suggested, A 1653 may even go back to Herodotus, who twice speaks of τῷ Αἰακῷ τέμενος in 5.89.3.

[18] In both of his books on Aigina, Welter (1938 and 1962) located the Aiginetan Aiakeion on Colonna Hill, below, i.e., south of, the southeast corner of the Temple of Apollo. Without giving any more information, he referred to a broad, ancient Propylon on the south side of the enclosure. In the museum at Aigina, inv. no. 752, is a fragment of an Archaic marble relief depicting a man in a chariot, which Welter identified as one of the reliefs mentioned by Pausanias decorating the walls of the entrance into the enclosure. It was found on Colonna Hill. For this sculpture see Walter-Karydi 1987, pp. 82–83, 126–128. In the preface to Wurster 1974 (p. 6), H. Walter argues, on the basis of Pausanias' text, that the Aiakeion could not have been on Colonna Hill; it was in the city, where the modern town of Aigina lies, on a hill to the east of the ancient naval harbor. Walter makes no reference to Welter's Propylon or to the finding-place of the relief. For this same view, see Zunker 1988, pp. 45–46, 69–72; Walter 1993, pp. 54–56; Walter-Karydi 1994, pp. 131–132.

A. E. Raubitschek has suggested to me, *per ep.*, that at Athens the temenos of the hero who saved the Greeks from famine (Paus. 2.29.7–8) may have been laid out near the old Λιμοῦ πεδίον, which seems to have been on the north slope of the Acropolis; see Judeich 1931, p. 298, note 2; Camp 1979, pp. 402–403. For Aiakos and the famine, see Zunker 1988, pp. 67–69.

Other brief literary references provide a few more helpful details about the Aiakeion on Aigina, which has as yet left no certain traces in the archaeological record. If they can be taken literally, Pindar's words in *N.* 5, lines 53–54, could evoke the entrance to the Aiakeion, προθύροισιν δ' Αἰακοῦ/ ἀνθέων ποιάεντα φέρε στεφανώματα σὺν ξανθαῖς Χάρισσιν ("carry to the front gates of Aiakos verdant crowns of flowers with the yellow-haired Charites"). For what it is worth, this is how the scholiasts interpreted these lines; see Drachmann III, page 100.

In *N.* 8, lines 13–15, Pindar may have in mind a particular statue of Aiakos, perhaps one that stood in his temenos in Aigina, ἱκέτας Αἰακοῦ σεμνῶν γονάτων πόλιός θ' ὑπὲρ φίλας/ ἀστῶν θ' ὑπὲρ τῶνδ' ἅπτομαι φέρων/ Λυδίαν μίτραν καναχηδὰ πεποικιλμέναν ("As a suppliant I clasp the hallowed knees of Aiakos, bringing to him on behalf of his beloved city and these his townfolk a variegated Lydian chaplet resounding with song"). We have seen above, pages 87–88, that the statues of the Aiakidai were probably kept in the Aiakeion and that a statue of Aiakos himself was regularly carried in procession from his temenos to the Sanctuary of Zeus Hellanios. Zunker (1988, page 70) properly asks whether or not these passages require a temple in the Aiakeion on Aigina.

If another brief reference to Αἰακιδῶν τ' εὐερκὲς ἄλσος ("the well-walled grove of the Aiakidai") in Pindar, *O.* 13.109, pertains to the Aiakeion, rather than to Aigina in general, we may have early evidence for the περίβολος τετράγωνος λευκοῦ λίθου and τοῦ περιβόλου δὲ ἐντὸς ἐλαῖαι πεφύκασιν ἐκ παλαιοῦ of Pausanias' description; cf. Zunker 1988, page 71.

The grain-tax law now permits us to go further. First, its two specific mentions of the Aiakeion in lines 14–16 clearly show that the sanctuary seen by Herodotus was still serviceable until at least 374/3 B.C. Second, in this temenos the Athenians are instructed to heap up a large quantity of grain (line 14). Unfortunately, the law does not specify exactly how much wheat and barley is in question, and our tentative estimates of at least ten-plus shiploads are based on flimsy data (see above, page 41). Third, the *nomos* instructs the *polis* to make sure that the Aiakeion is στέγον καὶ τεθυρωμένον (line 15), presumably to protect the grain from sun, rain, vermin, thieves, and unauthorized visitors. Recalling Herodotus' term τέμενος, we might infer from this passage in the law that Aiakos' large precinct was previously open to the sky and perhaps entered with little difficulty. The law now appears to call for a (new?) roof and door(s). Since the law was intended to go into effect without delay, it is possible that the roof and door(s) were not elaborate or permanent fixtures. Moreover, the renovated temenos must also have provided working space for the πριάμενοι and for the ten men elected annually to take care of the grain, weigh it out, test it for dampness and impurities, keep the barley separate from the wheat, and protect it during a period of at least three months while it was stored in the Aiakeion. Finally, there may be a suggestion in lines 41–42 that this sanctuary is in or near the Agora, since the ten men are to sell the state's grain at the appointed time ἐν τῆι ἀγορᾶι. This could mean that they sold it out of this new public granary itself and not necessarily in the grain market in the *asty*.[19] Herodotus, we recall, placed the Aiakeion ἐπὶ τῆς ἀγορῆς. See below, pages 94–95.

After this long review of the literary and epigraphical testimonia we are ready to search for an appropriate candidate for identification as the Aiakeion among the buildings excavated in the Agora. The temenos we seek must have been built before the Ionian Revolt. It may have suffered

[19] For the grain market in Athens, see Nikophon's law on silver coinage, *SEG* XXVI 72, lines 18–20: φαίνειν δὲ τὰ μὲν ἐν [τ]ῶι σί[τωι πρὸς] τὸς σιτοφύλακας, τὰ δὲ ἐν τῆι ἀγορᾶι κ[α]ὶ [ἐν τῶι ἄλ]λωι ἄστει πρὸς τοὺς τõ δήμο συλλογέ[ας] ("let denunciations concerning offenses in the grain market be made to the *sitophylakes;* those concerning offenses in the Agora and in the rest of the city are to be made to the *syllogeis* of the people"), with Stroud 1974, p. 180. This passage may imply that the grain market, ὁ σῖτος, in the *asty* was not strictly in the Agora, but cf. Theophrastos, *Characters* 3.3, οἱ πυροὶ ἐν τῇ ἀγορᾷ. For those who believe in an Alphitopolis Stoa in Athens, possibly outside the northwest corner of the Agora, on the street leading to the Dipylon, a logical place for the grain market in the *asty* might be near it; see Judeich 1931, pp. 364–365; Wycherley, *Agora* III, p. 193; *contra* Gallo 1984, p. 59. See also Gofas 1993, p. 169; Migeotte, forthcoming, "Ventes," note 25.

damage in the Persian destruction of Athens, but it was back in service at the time of Herodotus' visit. He saw it on the edge of the Agora and in that position it served as a topographic landmark in the Attic Stelai of 414 B.C., lying at the point where the deme of Kollytos bordered the Agora. It was still in existence in 374/3 B.C., at which time it must have been large, probably open to the sky and easy of access, but capable of being roofed and closed off by doors. If we can press the comparative evidence from Aigina, it might have been square or rectangular and built of worked stone. If we have interpreted *POxy* 2087 correctly, it served as Aiakos' Athenian abode, it stood near the Tholos, and *dikai* were written up there.

Before pointing to the structure in the Agora which best fits all these criteria, we can briefly reject the only competing candidate. The old topographers of Athens did not trouble themselves much with the location of the Aiakeion.[20] It seems not to have been until 1953 that any scholar associated the temenos with specific physical remains. In publishing a preliminary account of the excavation of a small Archaic *eschara* just south of the Peribolos of the Twelve Gods in the Agora, Thompson speculated that the hero to whom it was sacred might have been Aiakos.[21] We now see that the tiny scale of this monument, 1.76 × 3.77 m., rules it out of consideration. To my knowledge, no other candidates have been proposed.

A glance at the plan of the Athenian Agora in the 4th century B.C. (Fig. 5), will quickly reveal that the most plausible candidate for the Sanctuary of Aiakos among surviving buildings is the large enclosure in the southwest corner which in earlier publications has often carried the label "Heliaia."[22] In publishing the lawcourts of the Agora in *Agora* XXVIII, Boegehold describes this structure as the "Rectangular Peribolos." It is not my purpose here to retrace the long trail of identifications this venerable structure has traveled in the scholarly literature, from Heliaia to Gymnasium of Ptolemy to Theseion and back to Heliaia again. Contemplating its checkered modern history, we might shrink from imposing yet another label on these long-suffering ruins. Certainly, I do so in the knowledge that all my evidence is circumstantial. It is, however, better evidence than that used in support of all previous identifications of this structure. There has never been very persuasive archaeological evidence to suggest that it could have been a lawcourt.[23]

[20] E.g., Wachsmuth 1890, II, p. 424; Judeich 1931, p. 356.

[21] Thompson 1953, pp. 43–46. Wycherley (in *Agora* III, p. 49) found Aiakos "a likely candidate" for the deity worshiped here. Cf. *Agora* XIV, p. 132; followed by Travlos (1971, p. 3); Zschietzschmann (1973); Zunker (1988, p. 70). More prudent is Camp (1990, p. 97), who is willing to go only as far as "This type of altar is usually associated with a hero."

The only other object from the Agora Excavations known to me with possible ties to Aiakos is a fragment of an Attic red-figured krater, ca. 450–425 B.C., found southeast of the Altar of the Twelve Gods. On it is represented a Doric capital, with the name Αἰακός painted in white to the right, probably as a label for a now-lost figure; *Agora* XXX, p. 194, no. 344. For [ἥρωι Αἰ]ακῶι ἀνέθηκεν, misread on a base from the Agora Excavations, see Meritt 1963, p. 45, no. 61, with the correction on p. 438

[22] The structure south of the Tholos of unknown purpose and identity labeled "Strategeion" on plans of the Agora cannot have been the Aiakeion since: (1) although its outside dimensions, ca. 26 × 21 m., might satisfy the spatial requirements of the grain-tax law, it contains a series of small rooms which would have considerably reduced its capacity; (2) although there are said to be earlier remains below it, there is no indication that they are pre-Persian and the structure itself appears to have been built in the second half of the 5th century B.C.; (3) in its original form, as a roofed building, it hardly qualifies as a temenos; (4) it is probably not close enough to the Agora to have been cited as one of the landmarks of the house in Kollytos which bordered on both the Agora and the Aiakeion, *IG* I^3 426, lines 5–8. See Thompson 1948, pp. 167–169; 1955, pp. 54–55; *Agora* XIV, pp. 72–73. For the possibility that the Hero Strategos may have been worshiped here, see Lalonde in *Agora* XIX, pp. 9–10.

[23] Hansen (1989, pp. 232–235) brought powerful and conclusive arguments against identifying this enclosure as the Heliaia. In his judicious discussion of "structures that can be interpreted as lawcourts," Boegehold (*Agora* XXVIII, pp. 5–20) observes, "When . . . the full sum of indicators is displayed, exiguous as that sum may be, Building A [under the Stoa of Attalos] more likely than the Rectangular Peribolos was the Heliaia." But he still regards the Rectangular Peribolos as a lawcourt, known as Metiocheion and Meizon, which served 1,500 dikasts in "the '40's, '30's, and '20's of the 5th century" (pp. 12, 14).

Of all the numerous objects associated with legal activities excavated in the Agora, only one bronze ballot (out of 54) was found in or near the Rectangular Peribolos

This Rectangular Peribolos in fact meets very satisfactorily all the criteria for identification as the Aiakeion that we have culled from the literary and epigraphic sources. First, it is near the southwest corner of the Agora and the Tholos. It lies south of the boundary stones of the Agora in a position that aptly meets the requirements of Herodotus' phrase ἐπὶ τῆς ἀγορῆς, while being close enough for the ten men to sell grain out of it ἐν τῆι ἀγορᾶι. Its proximity to the Agora, while not being actually in it, also enables us to understand how the house in Kollytos in *IG* I^3 426 could be described as bordering on the Agora on one side and on the Aiakeion on the other. One possible location for such a house would be in the narrow tongue of land that extends northeastward right up to the boundary stones of the Agora between the Rectangular Peribolos and the Tholos. See Figure 5. Remains of private houses that were still standing when the property of the Hermokopidai was sold at auction have in fact been excavated in precisely this spot.[24] The most famous of these is the home of the philosophical cobbler, Simon. If one wished to define the position of a house lying a few meters to the south of Simon's, one might say that it borders on the Agora on one side and on the Rectangular Peribolos on the other. We have no record of the deme in which Simon's house was located, but Kollytos, as we have seen, looks like the most probable candidate.

The form of the Rectangular Peribolos also meets all of the criteria established by the literary and epigraphic testimonia. So far the excavators of this building have published their results in preliminary form only.[25] For convenience I quote first from the relevant section of Thompson and Wycherley's account in *Agora* XIV, pages 62–65:

> In its original form the structure was a walled enclosure, not quite a true rectangle in plan, with interior measurements of about 26.50 × 31 meters yielding an area of about 821 square meters. The principal entrance was in the middle of the north side; ... The enclosure wall stood on a stepped foundation across the north side, elsewhere on a rough packing of limestone ...; enough of the foundation remains in place to make the plan certain. The visible wall was built of squared blocks of *Aeginetan limestone* [my italics]. ... It was topped by a saddle-shaped crowning member with a projecting cornice on either side. On one side the soffit of the cornice was adorned with a hawksbeak bed moulding delicately painted. We have no clue to the height of the wall. It was presumably high enough to keep out intruders ... In the course of its long history, the Heliaia underwent many alterations. One of the first major changes is to be dated in the 4th century B.C., probably its third quarter. At that time a series of rooms was inserted in the west side of the original enclosure; they were provided with a light colonnade that faced on the remaining unroofed area. ... Within the 4th century, a water clock was installed against the north wall. ... The next major event ... involved the roofing of the remaining part of the old enclosure. ... This operation appears to have been carried out in the middle of the 2nd century B.C.

This account can now be supplemented by the detailed description of the remains contributed to *Agora* XXVIII, pages 99–103, by John McK. Camp II, who has restudied the architecture

(*Agora* XXVIII, p. 89, B 44 = B 1013). No date is given for its context. For its location, see *Agora* XXVIII, p. 63, fig. 4. One bronze dikastic pinakion (*ibid.*, p. 63, P 14 = B 1104) was found in the adjacent Southwest Fountain House. Another bronze ballot (*ibid.*, p. 87, B 7 = B 43), with a context of "4th, 3rd, 2nd centuries," said by M. Lang to have come "from the immediate vicinity of the large early structure for many years labelled Heliaia" (p. 85), was actually found more than 30 m. to the west of the Rectangular Peribolos on the other side of the road along the west side of the Agora. Examination of *Agora* XXVIII, fig. 4 strikingly demonstrates that the paucity of dikastic equipment found in and around the Rectangular Peribolos is in sharp contrast to the large concentration of findspots around the Tholos/Bouleuterion, the Stoa Poikile, and the remains of lawcourts under the Stoa of Attalos.

[24] For these houses see *Agora* XIV, pp. 173–177; Camp 1990, pp. 56–58.

[25] Thompson 1954, pp. 33–39; cf. Camp 1986, pp. 46–47; Camp 1990, pp. 180–181.

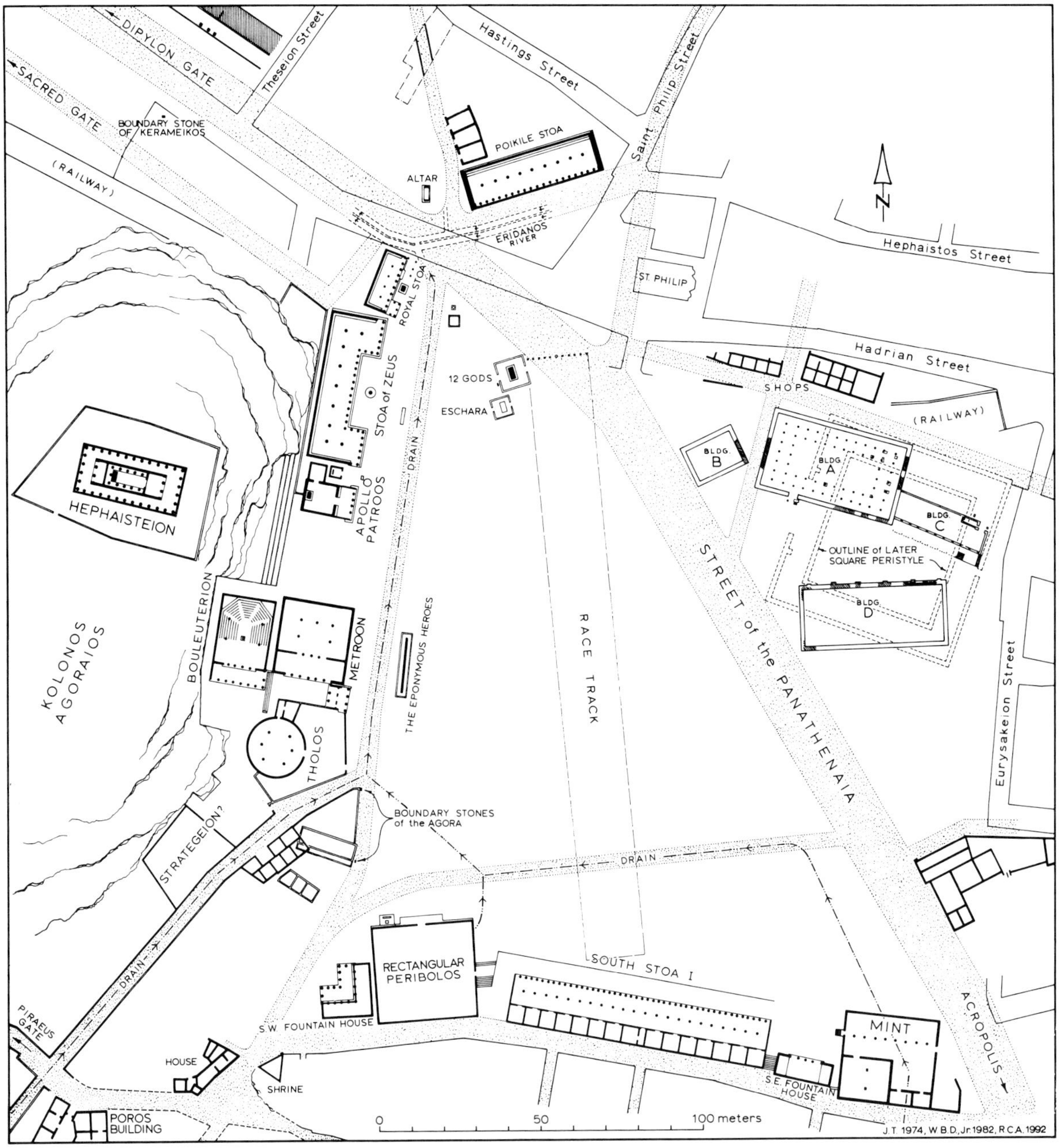

FIG. 5. The Athenian Agora, mid 4th century B.C.

of the Rectangular Peribolos and its associated pottery. While in essential agreement with the account of Thompson and Wycherley, Camp adds the following information pertinent to the present discussion. A *terminus post quem* for the original date of construction of the Rectangular Peribolos in the second quarter of the 6th century B.C. can be proposed on the basis of pottery excavated in two wells and in the remains of modest houses lying below the floor of the Peribolos. "The hawksbeak moulding on the soffit of the coping blocks finds its closest parallel in a raking geison dated to the time of the so-called Peisistratid temple on the Akropolis, sometime late in the second half of the 6th century" (page 100). The south wall was extensively rebuilt. "There is no sure way of dating this rebuilding, but because it [is] so extensive, it seems probable that it was occasioned by severe damage to the building, presumably in the Persian destruction of 480/79" (page 101). By at least ca. 340 B.C., several of the molded coping stones had been removed from the walls of the Rectangular Peribolos, for they were reused in the foundations of Building C, which was built about this time in the northeast sector of the Agora Excavations (R. Townsend, *Agora* XXVII, page 34). The addition of the rooms on the west side of the Peribolos or *possibly some earlier construction* necessitated their removal. The building was badly damaged in the Sullan sack of 86 B.C. and the area was thereafter used for industrial activity from the 1st century B.C. into the 2nd century A.C.—possibly as a pottery and bronze-working establishment. Noting the prominent location of the Rectangular Peribolos and its handsome construction, Camp supposed that it "was one of the major public buildings of Archaic and Classical Athens" (page 103). The absence of altars and appropriate small finds did not, however, encourage him to suggest a religious identification for the structure. Since "the basic ground plan . . . suggests a building intended to accommodate meetings or assemblies of some sort," he remained attracted to the identification of the Rectangular Peribolos as a court building, although he is careful throughout to avoid using the term *Heliaia*.

In proposing to identify this building as the Sanctuary of Aiakos, I would like to emphasize the following features. First, Herodotus' term τέμενος aptly describes this large precinct set off from the surrounding area by the peribolos wall. The size of this monument is also a determining factor in attempting to identify it as the Aiakeion of Agyrrhios' law, for any candidate must have the capacity to hold the anticipated quantities of the people's wheat and barley and to accommodate the activities of the πριάμενοι and of the ten men who weighed, looked after, and sold this commodity out of it. We can reach only a very rough estimate of how much space this grain required, because we must guess within only broad limits at the number of *merides* and shiploads of grain Agyrrhios had in mind. Using the comparative figures of the approximate, extrapolated yield of wheat and barley on Lemnos, Imbros, and Skyros in 329/8 B.C., we have estimated that the *dodekate* would at that time have produced a minimum of ca. 62 *merides*, or at least ca. 31,000 *medimnoi;* see above, pages 41–42. As noted, the totals contemporary with our law of forty years earlier may have been very different.

We do not know how this grain was shipped to the Peiraieus—either loose, poured in bulk into bins in the holds of transport ships, or, perhaps more likely, in sacks or amphoras.[26] The latter certainly would have been a more convenient method of conveying the grain up to the city, whether on pack animals or in carts, and of unloading it in the Aiakeion. How it was stored in the Aiakeion after the ten men had it weighed and tested it for dryness and impurities is also not stated. The verb κατανήσει in line 14, however, suggests that the πριάμενοι and the ten men may have heaped it up in bins where they stored it in bulk and not in sacks. There would, of course, have been separate compartments for wheat and for barley, and perhaps even individual storage areas for

[26] For transporting grain in bulk, in amphoras, and in sacks, see Casson 1971, p. 200; Velissaropoulos 1980, pp. 72–73; Rickman 1980, pp. 132–134; Gofas 1993, pp. 221–222. For sacks of chaff, see *IG* II^2 1672, lines 73–74, 108. See also the helpful study of Thompson (1983, pp. 64–75).

the produce of each of the three islands. We should expect variations in the quality of the wheat and barley from three different islands to be reflected later in higher or lower prices.[27] As far as we can tell from the form of ancient silos, grain stored in them seems normally to have been in bulk and not in sacks or jars.[28]

According to the calculations of Wallinga (1964, pages 23–24) and Lang (in *Agora* X, page 55, no. DM 70),[29] a *medimnos* of wheat would occupy ca. 0.50–0.52 m.³ A quantity of wheat and barley of ca. 31,000 *medimnoi* would therefore occupy ca. 31,000 × 0.52 = ca. 1,612 cubic m. The interior dimensions of the Rectangular Peribolos (26.50 × 31 = 821.50 sq. m.) are such that grain heaped up in it to a height of ca. 2.00 m. would occupy ca. 1,643 cubic m. Clearly, this figure would probably not leave enough space for one or more passageways giving access to the storage bins, nor for a working area for the ten men. In some of the bins, therefore, the grain may have been heaped up to a higher level than 2.00 m. Further speculation along these lines is probably futile since our total figure for the grain in question is so uncertain. It is clear, however, that the Rectangular Peribolos would have been roughly about the right size. Certainly no other single structure excavated in or on the edge of the Agora could have both qualified as a temenos and provided an equal amount of interior space in 374/3 B.C.

Two other formal characteristics of the Rectangular Peribolos support our proposed identification. First, it was in 374/3 B.C., as far as we can tell, still open and fairly easily adaptable for reuse as a granary. The existing western rooms, which would have considerably reduced the building's capacity for storing grain, were not constructed until well after the date of our law. The interior supports for a permanent roof in the building belong to a Late Hellenistic renovation. There is no indication that the Rectangular Peribolos was roofed before 374/3 B.C. or that it had any internal divisions. As we have seen, the height of the temenos wall, surmounted by its decorative coping stone, cannot be determined. Certainly in order to make the building dry, secure, and capable of holding large quantities of grain, the *polis* would have to have increased the height of the walls and added some internal supports for a roof covered by tiles. This might all have been easily carried out in wood, however, in relatively short order. Indeed the narrowness of the stone foundations, remarked upon by the excavators, makes walls of wood or mud brick more likely than those of stone. Dividing walls between wooden bins inside the structure might also have helped support the roof. We cannot hope to recreate the form of this temenos in its reincarnation as a public granary, but it is perhaps important to note that no surviving structural feature of the Rectangular Peribolos stands in the way of this hypothetical reconstruction of how the Aiakeion was rendered στέγον.

Second, in its first phase the Rectangular Peribolos had a broad stairway of five steps extending across the full length of its north side, which faced onto the Agora. It was thus very easy to enter. Even after the Propylon was added to the north side, there seems to have been little attempt to limit access, for its foundations are 12.10 m. wide.[30] Structurally, then, it would have been necessary to provide the Rectangular Peribolos with a secure door or series of doors, if large quantities of public grain were being heaped up in it. Again, wooden doors would probably have sufficed. They could have been added fairly quickly. We have no archaeological evidence for how this might have been effected, since nowhere has the top step on the north side survived *in situ*, but again it is not difficult to imagine how the Aiakeion could have been rendered τεθυρωμένον.

[27] Nothing is said in our law about a sample, or *deigma*, of grain which often accompanied cargoes of wheat and barley; see Velissaropoulos 1980, p. 73; Gofas 1993, p. 210. Perhaps all the *merides* were weighed and measured and since the ten men tested the grain for quality, a *deigma* was not necessary. For variations in price based on quality, see above, pp. 74–75.

[28] For possible storage of grain in the city of Athens, see Pritchett 1991, pp. 470–471. For storage of grain in large pithoi, sometimes set into the floor, see Amouretti 1979, pp. 60–61.

[29] See also Casson 1971, pp. 157–200.

[30] For this extension to the entrance, see Camp, *Agora* XXVIII, pp. 100–101.

Finally, by establishing the correct reading of line 18 of *POxy* 2087 as [ἐν δ(ὲ) or ἐπὶ] τῷ Αἰακίῳ δίκ(αι) ἀ(να)γράφον|ται (see above, pages 90–91), we can shed some light on another totally unexpected physical feature of the structure previously identified as the Heliaia. This is not the place for a detailed analysis of the exact meaning of δίκαι in this context on the papyrus, that is, whether it refers to notices of upcoming trials[31] or judgments/verdicts in trials already completed,[32] or something else. For us, the important term in trying to identify the Aiakeion is ἀ(να)γράφον|ται. Something legal is being written up, published, or recorded in, on, or at the Aiakeion. This verb could refer to documents on wood, papyrus, or even on stone that were stored or displayed in the Sanctuary of Aiakos. Cf. Andokides 1.83, ἀναγράφοντες ἐν σανίσιν ἐκτιθέντων πρὸς τοὺς ἐπωνύμους ("writing them up on boards, let them set them out at the Eponymoi"); and Aischines 3.39, ἀναγεγραφότας ἐν σανίσιν ἐκτιθέναι κελεύει πρόσθεν τῶν ἐπωνύμων ("he orders them to set them out in front of the Eponymoi after having written them up on boards"). That the verb, however, has its more literal, graphic meaning of "write up" as in ἀναγράψαι εἰς στήλην λιθίνην ("write up on a stone stele") or ἀναγράφειν εἰς τὸν τοῖχον ("write up on the wall),[33] is suggested by a remarkable archaeological discovery made in the excavation of this area. With the kind permission of Homer A. Thompson, who brought it to my attention, I first mention it here. In excavating a Hellenistic fill on April 29, 1953, just above bedrock immediately north of the water clock that was built against the north wall of the Rectangular Peribolos in the 4th century B.C.,[34] Vanderpool recovered more than fifty fragments of extremely high quality, smoothly polished, white wall plaster. Thompson (1954) did not include them in his excavation report (pages 33–39), but on the inventory card for these fragments of wall plaster in the Stoa of Attalos, A 3349, Vanderpool wrote in 1961, "Conceivably from north wall of Heliaia." He was referring to the front wall of the Rectangular Peribolos, the one that faced onto the Agora and for a time looked across at the nearby monument of the Eponymous Heroes, where public notices were prominently displayed. Camp (*Agora* XXVIII, page 100) noted that many of the surviving limestone wall blocks of the Rectangular Peribolos showed signs of "having been plastered on one face with a thin layer of fine marble-dust stucco."

Now the fascinating thing about these fragments of white wall plaster is that most of them carry letters written in a thin red paint. The letters are large, ca. 0.065 m. tall, with strokes ca. 0.01 m. wide. They are almost monumental in appearance and were clearly meant to be read from a distance. The letters are fairly carefully shaped and do not give the impression of belonging to casual messages or scribblings.[35] Unfortunately, the letters are so large and the fragments of wall plaster so badly broken that no words can be made out. The longest string of letters occurs on the piece illustrated in Figure 6, [– – –]ΤΡΑΤ̣[– – –], which measures ca. 0.14 × 0.17 m. Many other fragments bear traces of single letters. Dating by letter-forms—always a hazardous exercise—is of little assistance here, but the letters are not ornate or decorative. The crossbar on the alpha

[31] Cf. Xenophon, *Apologia* 1, ἐπειδὴ ἐκλήθη εἰς τὴν δίκην ("when he had been summoned to the trial"); Demosthenes 21.103, πλὴν ἵν' ἐκκέοιτο πρὸ τῶν ἐπωνύμων καὶ πάντες ὁρῷεν, «Εὐκτήμων Λουσιεὺς ἐγράψατο Δημοσθένην Παιανιέα λιποταξίου» ("except in order to have posted up in front of the Eponymoi, so that all could see, 'Euktemon of Lousoi has indicted Demosthenes of Paiania for desertion' "); Libanios, *Oratio* 29.43, τῶν πρὸ τῶν ἐπωνύμων ἐκκειμένων ὁ δεῖνα τοῦ δεῖνα ἐγράψατο; and Isaios 5.38.

For the publication of notices of trials, see Harrison 1971, pp. 91–92; *Syll.*³ 344, lines 35–40.

[32] For the publication of verdicts, see Aristotle, *Politics* 6.1321b34, ἑτέρα δ' ἀρχὴ πρὸς ἣν ἀναγράφεσθαι δεῖ τά τε ἴδια συμβόλαια καὶ τὰς κρίσεις [ἐκ] τῶν δικαστηρίων ("there is another magistracy before whom it is necessary to have written up private contracts and decisions from the lawcourts"). For the rarity of the procedure at Athens, see the pertinent remarks of Boegehold (*Agora* XXVIII, pp. 7–8, 23–24, 30–31), who collects a useful list of examples on pp. 177, no. 148; 237, no. 342a.

[33] For the latter, see Andokides 1.84; *IG* I³ 84, lines 24–25.

[34] For this water clock, see Camp 1986, pp. 157–159; 1990, p. 181.

[35] Such as erotic graffiti of the type familiar from Aristophanes, *Acharnians* 144, with scholia and the good note of Rennie 1909, *ad loc.*

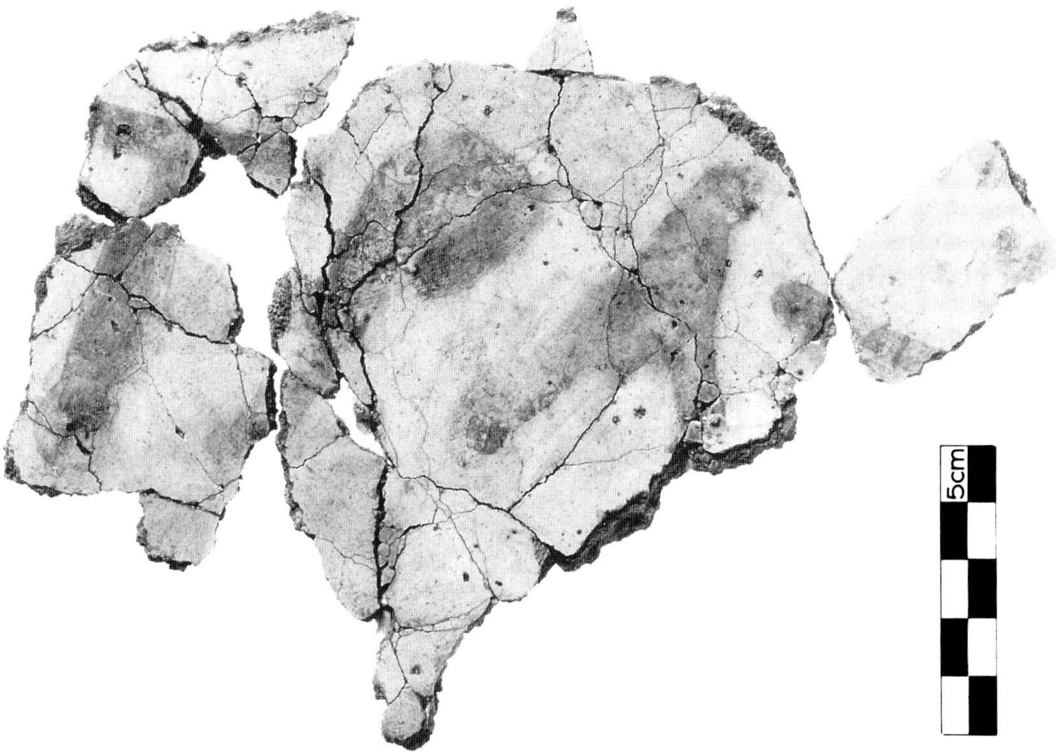

FIG. 6. Plaster fragment with painted letters, inv. no. A 3349

is horizontal, not broken. There are no serifs or other obvious characteristics that would lead one to place them later than the Classical period. It may also be significant that they were excavated in a fill of Hellenistic times. Probably by this time the inscribed plaster had fallen or been removed from its original position on the wall, perhaps on the north wall of the Rectangular Peribolos.

There is of course no way, on present evidence, that we could claim that these letters recorded δίκαι, whatever that term means in *POxy* 2087, but if, as seems likely, the plaster once adhered to the exterior of the north wall of the Rectangular Peribolos, then we have perhaps another suggestive archaeological link between this building and the literary testimonia. The latter tell us that the Aiakeion was used for the "writing up" of *dikai*, and we now have good reason to believe that the front wall of our proposed candidate carried painted inscriptions that could be erased and replaced on its polished plaster surface by other, more current messages.[36]

Rutherford has aptly pointed out to me *per ep.* the suitability of posting δίκαι on the front wall of the temenos of Aiakos, the hero ὃ καὶ δαιμόνεσσι δίκας ἐπείραινε ("who even for the gods settled *dikai*") (Pindar, *I.* 8, lines 24–25). Cf. also Plato, *Apology* 41A; *Gorgias* 523E–524A; Isokrates 9.15; *LIMC* I.1, page 311, nos. 1–3. See also Zunker 1988, pages 78–80; 86–88; Carnes 1995,

[36] I am indebted to John Camp for examining these fragments with me, for much helpful discussion, and for providing the photograph printed in Figure 6.

pages 21–24. It is not out of the question that trials might have been held in this temenos before 375/4 B.C. without leaving any trace in surviving ancient literature.

There might be some attraction to bringing δίκ(αι) ἀ(να)γράφον|ται at the Aiakeion in *POxy* 2087 and the inscribed front wall of the Rectangular Peribolos into play with Aristophanes, *Knights* 977–980:

καίτοι πρεσβυτέρων τινῶν
οἵων ἀργαλεωτάτων
ἐν τῷ δείγματι τῶν δικῶν
ἤκουσ' ἀντιλεγόντων

And yet I heard some old men
of that most contentious kind
arguing at the *Dikai* Bazaar.

The scholiasts on this passage are surely wrong to interpret δεῖγμα literally as the well-known area in the Peiraieus where merchants spread out samples of their wares. The contentious old men (*dikastai*?) are arguing at a place in Athens where *dikai* were metaphorically on display like goods for sale. It is more likely that δεῖγμα here strikes this comic note, as editors such as Ribbeck, Neil, Van Leeuwen, and Sommerstein have observed.

We have seen (above, page 99) that the Monument of the Eponymous Heroes was such a place and indeed Wycherley (*Agora* III, page 86, no. 231) and Shear (1970, pages 203–204, 219) have associated Aristophanes, *Knights* 977–980 with this structure. *POxy* 2087 now shows that *dikai* were also on display at the Aiakeion and the inscribed plaster fragments suggest the possibility that the Rectangular Peribolos could have had *dikai* painted on its front (north) wall. The likelihood that this building may in fact have been close to the spot the chorus of the *Knights* jokingly designated as ἐν τῷ δείγματι τῶν δικῶν grows greater when we remember that at the time of the play, 424 B.C., the statues of the Eponymoi stood only a few steps north of the front wall of the Rectangular Peribolos. See Shear 1970, pages 203–222, especially page 205, "The long and narrow rectangle of the monument was oriented with its axis almost exactly north and south and surveyed in obvious relation to the northwest corner of the Heliaia, from which it was separated by an interval of 8 m." If, as Shear has argued, the earliest phase of the Monument of the Eponymous Heroes was not built until ca. 430 B.C., the front wall of the Aiakeion might have served as the "*Dikai* Bazaar" before that date. John Camp reminds me that the Δεῖγμα in the Peiraieus was, like the Rectangular Peribolos, a building. For the testimonia, see Gofas 1993, pages 79–85.

It is remarkable how closely this temenos on the edge of the Athenian Agora resembles Pausanias' description of the Aiakeion on Aigina in location (ἐν ἐπιφανεστάτῳ δὲ τῆς πόλεως), form (περίβολος τετράγωνος λευκοῦ λίθου), and size (τοῦ περιβόλου δὲ ἐντὸς ἐλαῖαι πεφύκασιν ἐκ παλαιοῦ) (2.29.6–8). There is a further possible connection with the island of Aiakos, a tantalizing one that I leave to those who are competent to judge such matters. Without making any connection between this island and what they then regarded as the "Heliaia," Thompson, Wycherley, and Camp have all described the material of the peribolos wall as "Aeginetan limestone" or "poros."[37] If in fact the basic material of the earliest phase of construction of the temenos was imported from quarries on Aigina, comparison of the two shrines of Aiakos gains in importance.

Apart from its form, there is no other archaeological evidence to indicate that the Rectangular Peribolos was a sanctuary. The excavators found no deposits of votives, no inscribed dedications, no remains of an altar or the like. The interior of the building, however, was very poorly preserved

[37] *Agora* XIV, p. 62; Wycherley 1978, p. 35; Camp 1990, pp. 180–181. In *Agora* XXVIII, pp. 99–101, Camp, perhaps more cautiously, calls the material "hard, gray limestone," avoiding the term *Aiginetan*. For Aiginetan stone in Athens in 409/8 B.C., see, e.g., *IG* I^3 475, line 13; at Eleusis in 408/7 B.C., see *IG* I^3 386, lines 92–95; Mark 1993, p. 42.

and ravaged by later construction and disturbances. It may be that originally Aiakos had no more than a simple altar, one that could not be expected to leave much evidence behind. Pausanias notes that the altar in his peribolos on Aigina, which according to some served as his tomb, projected only a little above ground level; in other words, it was not a substantial construction.

Finally, the Rectangular Peribolos in the southwest corner of the Agora also meets all of our chronological criteria. It was certainly built before the time of the Ionian revolt. I do not know if it is possible to be more precise about its date of construction than Camp's suggestion of the second half of, or late, 6th century B.C. based upon the pottery found beneath it and the profile of the molding on the coping stone of the wall. But on present evidence nothing seems to exclude a construction date that would accord with Herodotus, *viz.*, ca. 506–499 B.C. Although it may have been damaged in the Persian destruction of Athens, repairs to the south and west walls and construction of the Propylon demonstrate that the building was in operation at the time of the historian's visit and available for storage of large quantities of grain in 374/3 B.C.

We hear no more of the Aiakeion after the grain-tax law (except for the lexicographical notes), and we do not know how long the practice of storing public grain in it continued. This may, of course, simply be a function of the fragmentary state of our literary and epigraphical sources. It is always dangerous to build sand castles on silences in our testimonia, especially on those of Pausanias. His failure to note the existence of our building, although he enumerates other Athenian lawcourts, has always been an obstacle to those seeking to identify it as the Heliaia. Even more potentially awkward for us might be his silence about the Aiakeion in Athens, particularly since he was so well informed about the parent shrine in Aigina. It is possible, however, that the massive secular and commercial functions imposed on this structure by Agyrrhios' law marked an important stage in transforming it from a religious to a more utilitarian building. Possibly the rooms added in the second half of the 4th century B.C. represent the next stage in the process. The water clock built in front of it might point in the same direction. The Hellenistic colonnade and possibly the permanent roof described by Thompson and Wycherley may also have contributed to the transition. In the Sullan sack of Athens, 86 B.C., much damage was inflicted on this building. Thereafter it seems to have been the site of assorted industrial activities. If by the time of his visit, the venerable peribolos of Aiakos was no longer recognizable in this corner of the Agora, the silence of Pausanias becomes explicable.[38]

Our law sheds no light on the question of the Aiakeion's sacred status at the time when it was structurally transformed into a public granary. The hero who had once saved the world from famine may have been judged a suitable patron for the new function his temenos now served, but the law of 374/3 B.C. could also indicate that by this date worship of Aiakos had here virtually come to an end.

Before leaving the Aiakeion, we need to consider briefly the remarkable theory proposed by Oikonomides (1990) that there was below the Tholos a "house of Aiakos which existed on the same site for almost 300 years . . . The precise site in Melite, where the 'house of Aiakos' was established by his grandson [*sic*] Eurysakes, is now under the foundations of the Tholos, and his family burial

[38] For the damage to the building in 86 B.C. and later, see Camp, *Agora* XXVIII, p. 102. Lalonde, *Agora* XIX, p. 9, finds "a decline in devotion to heroes in the Hellenistic age" in the Athenian Agora, represented inter alia by the destruction of the rectangular heroön near our candidate for the Aiakeion at the time of construction of the commercial Middle Stoa about 180 B.C.

While eulogizing Euagoras of Salamis, who died in the year of Agyrrhios' law, Isokrates (9.15) mentions the Cypriote king's ancestor Aiakos on Aigina but says nothing of this hero's large shrine in Athens, although the Athenians had honored Euagoras with a statue in the Agora. By the time of the speech, the Athenian Aiakeion may have already been transformed into a public granary; for speculation on the (unknown) date of the *Euagoras*, see Shrimpton 1987. Race (1987) discusses Isokrates' use of the Aiakos legends in this speech.

Kroll (*Agora* XXVI, p. 295) conjectures that in the middle of the 3rd century A.C. the building served as the mint for ancient Athens' last coinage, the Period VI imperials of A.D. 264–267.

plot is nearby . . . The last burials in the family plot date from the end of the sixth century B.C., so it is evident that there were no more descendants. The old 'house of Aiakos' must still have been standing on its site in Melite, facing the Pelargikon ring-wall and the Acropolis hill." In response to the Delphic Oracle of "ca. 488," the Athenians laid out "the 'Aiakeion' sacred precinct . . . close to the old house where Aiakos' descendants had lived on the outskirts of the city for three centuries . . . The 'Aiakeion' and the old house stood there until the evacuation of Athens in 480/79 B.C. and the arrival of the Persians. From the excavations we know that they were then destroyed by fire."

I have quoted at length because no paraphrase could adequately bring out the confusion this theory imposes on both the literary testimonia and the archaeological remains. Oikonomides was led to propose this reconstruction partly on the basis of his restoration and interpretation of the lexikon entry in *POxy* 2087, lines 16–18. I have attempted to show elsewhere (Stroud 1994) that the connection he tries to make between Aiakos and the Tholos is based on faulty readings and impossible syntax; it is without foundation in the ancient textual evidence. Here we need to consider two further weaknesses in this theory: one legendary, the other archaeological.

First, when Eurysakes left Salamis to take up residence in Melite, no surviving ancient source says that he brought his great-grandfather Aiakos with him. In fact, his own father Ajax, as we have seen, retained his own sanctuary, the Aianteion, on Salamis. In all the substantial literary and epigraphic testimonia on Eurysakes and the Eurysakeion, conveniently collected in *Agora* III, nos. 246–255, there is no mention of Aiakos in Athens, let alone a residency of 300 years in a "house" built by his great-grandson. It is not until the time of the Delphic Oracle of ca. 506 B.C. that we hear of Aiakos having a sanctuary in Athens. Herodotus 5.89 implies that the Aiginetan hero was first brought over to live in Athens at that time, not that he had been living there for several centuries and was now for the first time given a temenos near his "house." In fact, there is no evidence for a "house" of Aiakos at all; Oikonomides simply inferred its existence from οὗ φασι τὸν Αἰακὸν οἰκῆσαι in *POxy* 2087, lines 16–17 and Hesychios. But the only place where the testimonia say that the hero "lived" is in his temenos, the Aiakeion, which was not laid out until ca. 506 B.C.[39]

Second, despite Oikonomides' claim (1990) that his theory demonstrates the triumph of "logic and simplicity" over "forced and unrealistic identifications" (that is, those of Thompson and Wycherley), he has seriously misrepresented the archaeological remains below the Tholos. Without citing the source, he strings together a series of quotations from *The Athenian Agora: A Guide to the Excavation and Museum* (4th ed., Athens 1990, pages 53–54) to produce the following paragraph:

> The text [*POxy* 2087] leads the reader to assume that there exists under the foundations of the Tholos exactly what the excavations revealed: a large house "*made up of a colonnaded court surrounded by rooms of various shapes and sizes. The building dates from about the middle of the 6th century* B.C." How important this house was is obvious from the fact that "*the foundations, made of rubble stonework, have been largely covered over again to assure their preservation.*" Attached to the house "*was encountered a*

[39] Wycherley (*Agora* III, p. 48) urged, against the verb οἰκῆσαι in Hesychios, that "there is no question of Aiakos *living* at Athens," but the metaphor of a hero "living" in his own sanctuary is familiar from Pindar, *N.* 7.47, where the Aiakid Neoptolemos was fated to οἰκεῖν in Delphi. See also Aristophanes, *Wasps* 389–391, ὦ Λύκε δέσποτα, γείτων ἥρως . . . ᾤκησας γοῦν ἐπίτηδες ἰὼν ἐνταῦθ' ἵνα ταῦτ' ἀκροῴο "(O master Lykos my neighboring hero . . . you have come to live here so conveniently as to hear these sounds"), with scholia and MacDowell's note *ad loc.*; cf. Boegehold 1967; *Agora* XXVIII, pp. 188–189. (I am indebted to Jorge Bravo for this apt reference.) Cf. also the play on the οἶκος/τέμενος of the hero Protesilaos in Herodotus 9.116. Inscribed on doorways, the metaphor lived on for centuries in pagan and Christian religion as ὁ τοῦ Διὸς παῖς καλλίνικος Ἡρακλῆς ἐνθάδε κατοικεῖ ("The son of Zeus, Herakles, glorious in victory lives here") and Χριστὸς ἐνθάδε κατοικεῖ ("Christ dwells here"). Among many other discussions of this formula, with bibliography, see Rusten 1983; Merkelbach 1991; Faraone 1992, pp. 58–59. I thank Kathryn A. Morgan, Deborah Boedeker, and William Morison for helpful advice on this point.

> *family burial plot containing 22 graves of the 8th, 7th, and early 6th centuries* B.C." So what the text says is absolutely verified by the excavation finds. [italics Oikonomides]

If, however, we read *POxy* 2087 the way Oikonomides wants us to, surely we do not expect to find below the foundations of the Tholos a "house of Aiakos" built as late as ca. 550 B.C. We are asked to believe that he lived in this "house" for "almost 300 years." Eurysakes then must surely have established this alleged abode of his great-grandfather in Melite much earlier than the middle of the 6th century B.C. Sensing this chronological conflict between the excavated remains and his theory, Oikonomides resorted to the family burial plot. It is important to note that the words "attached to the house" are his, not those of the *Agora Guide*. In fact, the latter describes the house under the Tholos, Building F, and the "Early Cemetery South of the Tholos" in two separate sections without stating or implying any physical or historical connection between the two (pages 53–54). The latest burials in the cemetery were early 6th century B.C., not "the end of the sixth century B.C.," as Oikonomides (1990, page 22) states. The grave precinct also lay at a much deeper level than the house, which was only built later. It is Oikonomides who has made the connection between the two, not the "excavation finds," which provide no plausible evidence for combining later house and earlier cemetery.

Finally, in his zeal to expose the "forced and unrealistic identifications" of the American excavators, Oikonomides argues that the importance of the house in antiquity "is obvious from the fact that '*the foundations, made of rubble stonework, have been largely covered over again to assure their preservation.*'" Not only has he slightly misquoted the *Agora Guide* here, but he has also failed to grasp the meaning of its straightforward English: "The rubble stonework foundations have been largely covered over again to assure their preservation; their tops project above the ground to the south of the Tholos" (page 54). This means, of course, that it was the American excavators, not the ancient Athenians, who covered over these foundations.

I have devoted so much space to an examination of Oikonomides' theory not merely because it was presented in his own journal in offensive, polemical language that casts unwarranted doubts on the scholarly integrity of Thompson and Wycherley,[40] nor because it provides an instructive example of faulty method in drawing inferences from a papyrus text, from legends, and from the results of excavation. My aim rather is to place some serious obstacles in the path of others who might be tempted to believe in a "house of Aiakos" under the Tholos on the basis of *POxy* 2087. This theory should be allowed to die the quiet death it deserves, but, scholarship being what it is, I fear that we may not have seen the end of the "house of Aiakos."

I do not believe that Flavius Philostratos' brief reference to the transfer of the *oikos* of the Aiakidai from Aigina to Salamis before the battle of 480 B.C. can be brought into play with Oikonomides' theories about a house of Aiakos in the Athenian Agora (*Heroikos* 53.15). See the good discussion of Zunker (1988, page 72).

TRANSPORTATION OF THE GRAIN FROM THE PEIRAIEUS TO THE AIAKEION

If the temenos of Aiakos was located in the southwest corner of the Agora, how did the tax-collectors or their representatives reach their destination in transporting wheat and barley up from the Peiraieus? The two most logical routes from the large commercial harbor in the Peiraieus (Κάνθαρος) would have been (1) the *hamaxitos* road, located outside the north Long Wall,[41] and (2) the road that was framed by the two Long Walls themselves. The latter, while

[40] See Stroud 1994, p. 1, note 2.

[41] For the *hamaxitos*, see Xenophon, *HG* 2.4.10; Plato, *Republic* 439E. Pausanias (1.2.2) probably followed this road up from the Peiraieus, for he sees the ruined walls of Athens and begins his account of the city at the Dipylon. This seems also to have been the route followed by

more direct, is also steeper, for it had to climb the rocky hills of southwest Athens to enter the *asty* through one of the gates to the north or south of the Pnyx.[42] The *hamaxitos*, seeking level terrain, approached the *asty* in its northwest sector and thus avoided the rocky barrier of the Mouseion, Pnyx, and the Hill of the Nymphs. From this road one could turn off to enter the *asty* through either the Demian Gate or the Peiraieus Gate, or, much farther north, through the Sacred Gate or the Dipylon.[43] The advantage of this route for πριάμενοι conveying grain to the Aiakeion in carts is that it avoids a climb up into the city and is not all that much longer than the more direct route through the gap between the Mouseion and the Pnyx. The latter would presumably have been the road used by pack animals. See Figure 7.

Entry through the Demian Gate[44] or the Peiraieus Gate[45] would have been more attractive than using the Sacred Gate or the Dipylon, if I am right in placing the Aiakeion in the southwest corner of the Agora. Heavily laden carts would not only thereby have a shorter, more direct route to their destination, but they would also avoid having to make their way across a busy and perhaps congested market square, either on the road leading south from the northwest corner of the Agora past the civic buildings and the Tholos or on the Panathenaic Way. There are no traces of wheel-ruts on the former and Young (1951, pages 166–168) suggested that wheeled traffic entering the city through the Dipylon and headed for the Agora might have followed his

---

Lucian's fictitious interlocutors who walk from the Peiraieus to the Dipylon in his *The Ship; or, The Wishes* (Judeich 1931, pp. 139–140, 186; Travlos 1971, p. 164).

[42] For this road, see Judeich 1931, pp. 180, 186, who identified it as ἡ διὰ Κοίλης ὁδός, Herodotus 6.102; Thompson and Scranton 1943, pp. 312–315; Hill 1953, p. 201; Travlos 1960, p. 198; 1971, p. 169, nos. XIV, XV (gates); 1988, p. 34; Lauter 1982, pp. 44–52. Its northern section, extending into the Agora, was called "Areopagus Street" by Young (1951, pp. 145–149).

On the topic of these roads, I have consulted with profit Conwell 1992, a copy of which is in the Blegen Library of the American School of Classical Studies at Athens. We still await a detailed study of the roads and streets of ancient Athens. Masses of primary data are lying untapped in the excavation reports of the Greek Archaeological Service published annually in Δελτ, excellent fodder for Ph.D. dissertations.

[43] For the location of these four gates in the west city wall, see Travlos 1971, p. 161, nos. I, II, III, IV; 1988, p. 34.

[44] For the Demian ("Executioner's") Gate, see Plato, *Republic* 439E; Plutarch, *Themistokles* 22; Hesychios, *s.v.* Δημίασι Πύλαις; Judeich 1931, pp. 140, 186; Travlos 1960, p. 52; 1971, pp. 121, 159, fig. 219, I; 1988, p. 34, I. Apparently no trace of this gate has yet been reported. Judeich and Travlos' hypothetical placement of it near the intersection of Akamantos and Aktaiou Streets accords with the orientation both of the wheel-ruts reported by the former near the intersection of Akamantos and Erysichthonos Streets to the west (Judeich, plan I) and of the road excavated by Threpsiades in 1958 immediately west of the Sanctuary of Artemis Aristoboule (Threpsiades and Vanderpool 1964; Travlos 1971, p. 121). A gate here would lie conveniently close to and immediately north of the steep, rocky Hill of the Nymphs. To locate the Demian Gate (and the fortification wall) farther to the northwest, down the hill, would be to lose this important topographical advantage. Judeich's wheel-ruts, therefore, will have belonged to the road outside, to the west of the Demian Gate. The gate may itself have led only to the place of execution, or *barathron*. One disadvantage for carts coming from the Peiraieus and turning off the *hamaxitos* at this point to climb up to the probable location of the Demian Gate is that the approach today is quite steep along any of the three modern streets leading to the intersection of Akamantos and Aktaiou, that is, Odos Kymaion (and its western extension, Thorikion), Akamantos itself, and Galateias. Immediately north of Odos Galateias, however, along the hypothetical line of the city wall, there is a significant drop in elevation down to the level and relatively low-lying environs of the Peiraieus Gate, ca. 250 m. to the north. Also, the story of Leontios gazing at corpses lying here under the city wall may indicate that this was not a very attractive route (Plato, *loc. cit.*).

[45] For the Peiraieus Gate, see Plutarch, *Sulla* 14; Judeich 1931, pp. 139–140, 186; Travlos 1960, p. 198; 1971, p. 159, fig. 219, II; 1988, pp. 23, 36. For the excavated remains of this gate with a section of the Themistoklean city wall, the *proteichisma*, the *taphros*, a tower of Justinian, and part of a road running parallel to the wall, see Threpsiades 1953, pp. 61–71; Philippaki 1966; Spathari 1982; Lygouri-Tolia 1985a; 1985b; 1988. The massive reference to Noack 1907, pp. 123–160, 437–566, repeatedly cited in these last four publications as throwing light on the Peiraieus Gate, can be reduced to the only relevant pages, pp. 500–508, where he discusses a round tower on the circuit wall between the Dipylon and the Hill of the Nymphs.

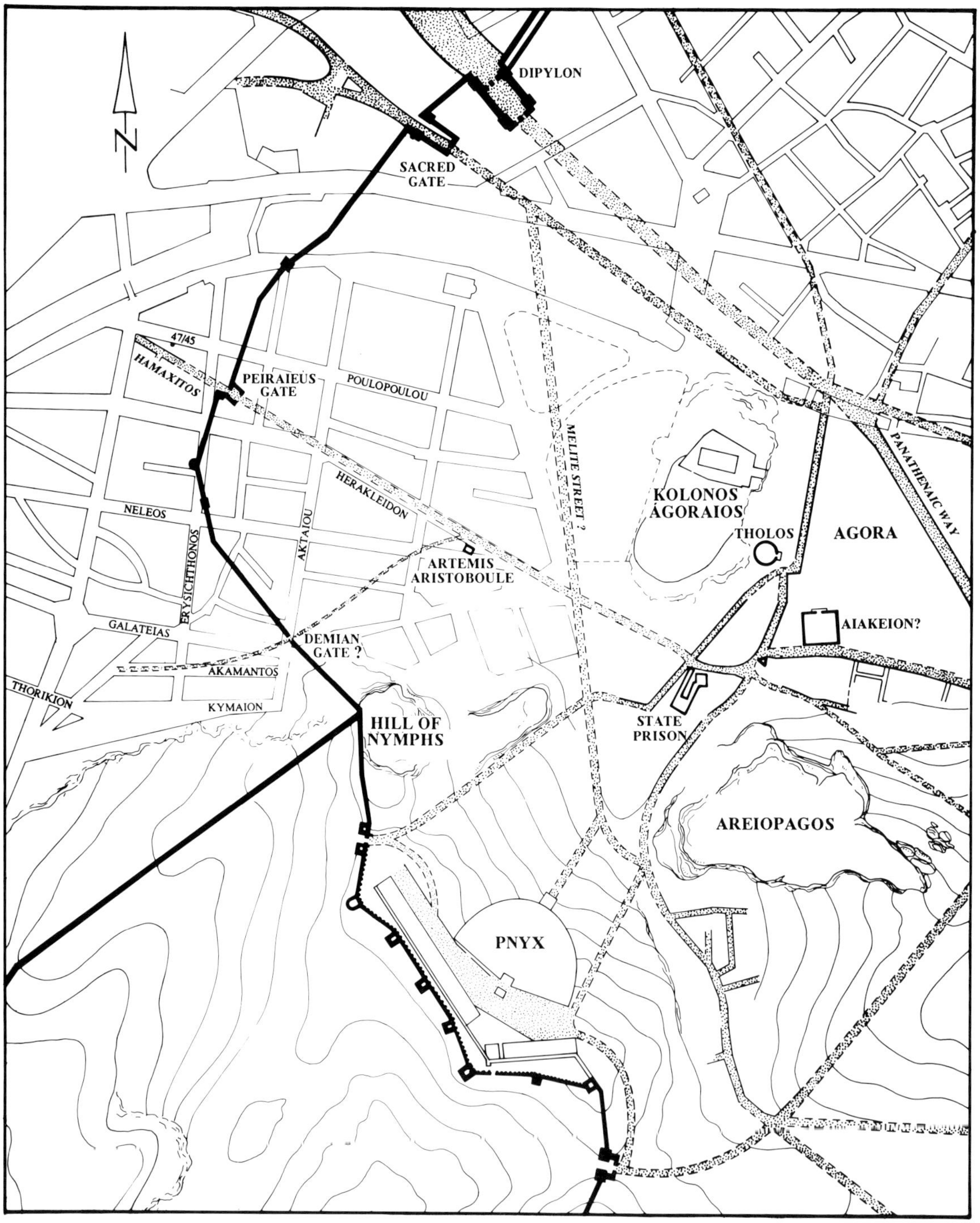

FIG. 7. Northwestern Athens

"Melite Street," which probably lay well to the west of the market square.[46] See his plan on page 146, with discussion on pages 167–168, and Camp 1986, page 182.

Recent excavations now make it possible to plot the course of the ancient road that entered the *asty* through the Peiraieus Gate and continued eastward to the Agora. Alexandri of the Greek Archaeological Service established its orientation on a northwest/southeast line outside the walls by the discovery of a stretch of ca. 35 m. of this ancient road under what is now the large apartment building at 45–47 Odos Poulopoulou,[47] about 40 m. west of the Peiraieus Gate. The oblique angle at which the (preexisting?) line of this road meets the fortification wall probably accounts for the fact that the excavators have found that the Peiraieus Gate itself is not perpendicular to the wall but was set at the same angle as the road.[48] When the line of this road is extended hypothetically through the remains of the Peiraieus Gate toward the southeast, it strikingly conforms exactly with the course of the modern Herakleidon Street beyond its intersection with Aktaiou Street. See Figure 7 and Travlos 1988, page 34. This orientation also conforms neatly with that of the north wall of the Temple of Artemis Aristoboule excavated by Threpsiades in 1958 at the intersection of Herakleidon and Neleos Streets.[49] Presumably the temple lay close to the ancient road on its south side.[50] Another possible indication that Herakleidon Street might mark the line of an ancient road, as it proceeds southeastward from Aktaion Street toward the Agora, is the fact that it is the only street to cut obliquely across all the other streets in this sector of Athens, which are oriented roughly north–south or east–west.

Beyond the Temple of Artemis Aristoboule, continuing the line of modern Herakleidon Street, there is excellent excavated evidence for where this important artery in the road system of ancient Athens approached the Agora. The latter was the target of our *priamenoi* conveying wheat and barley from the Peiraieus to the *asty* for storage in the Aiakeion and subsequent sale ἐν τῆι ἀγορᾶι, lines 41–42. Extensive remains of this road were revealed in the Agora Excavations of 1947 and 1948, at the north end of the "industrial district" outside the southwest corner of the market square. Survival of a stone bridge permitted the excavators to pinpoint the place where our road, which they called "Piraeus Street," crossed the Great Drain.[51] This crossing is only ca. 100 m. south of the Tholos and equidistant from the Rectangular Peribolos. Direct access to the latter was easy for carts once they had crossed the bridge, for the road then continues to the east, passing north of the Poros Building, or "Prison." It then turns north between the remains of a Classical house and the triangular *hieron*. From this point the road branches both to the north and to the east. The former branch skirts "the northwest corner of the great square building which has been tentatively identified as the Heliaia, and it opened into the central area of the Agora in front of the Tholos" (Shear 1970, page 205). Here, excavations in 1967 revealed extensive remains of wheel-ruts in the road surface which show that carts could and did reach the front of the Aiakeion in antiquity. The eastern branch of the road passed immediately to the south of the Rectangular Peribolos at a higher level, also giving access to this side of the building for carts.

[46] Young's further suggestion, *loc. cit.*, that the Agora was closed to wheeled traffic is difficult to square with Pollux 7.192, ἐπίπρασκον δ' οὗτοι τὸ γλεῦκος ἐφ' ἁμαξῶν εἰς τὴν ἀγορὰν κομίζοντες ("these men used to sell sweet wine from wagons bringing it into the Agora"), and the wheel-ruts east of the Tholos to be described presently.

[47] For these excavations see ΔελτΔ 24, 1969 [1970] Β′, pp. 64–68.

[48] See the very helpful plan of Lygouri-Tolia 1985, p. 18 and note 45 above.

[49] See above, note 44.

[50] For another(?) road leading from the Sanctuary of Herakles Alexikakos in Melite to the Agora, see *IG* II^2 1582 + *Hesperia* 5, 1936, pp. 393–413, no. 10 + *Agora* XIX, P26, lines 453–454; Wycherley 1959, pp. 67–68; Woodford 1971, pp. 218–219.

[51] Young 1951, pp. 151–160.

THE SALE OF THE GRAIN IN THE AGORA

If the identification of the "Heliaia"/Rectangular Peribolos as the Aiakeion is correct, a plausible venue for the sale of the grain would have been the fairly large, open area extending to the north of the existing foundations of the archaic temenos. The two nearby boundary stones of the Agora still *in situ* (*IG* I^3 1087, 1088) show that temporary booths or stalls set out here for the sale of wheat and barley from the islands would have been located ἐν τῆι ἀγορᾶι. See Figure 5. This location has the additional advantage of proximity to the Aiakeion, thus reducing the distance that sacks of grain would have to be carried.[52] Also, before the construction of the Hellenistic Middle Stoa, this area might have been free enough of buildings to accommodate the fairly large number of buyers that we should probably expect to turn out for a sale of the people's grain. In preparation for consolidation of the foundations of the western rooms of the Hellenistic Middle Stoa, John Camp and his excavators have exposed in 1996 and 1997 a large area in exactly this region. It has remained free from structures that are contemporary with Agyrrhios' law of 374/3 B.C. We noted above, page 93, that it is probably significant that the sale of the people's *sitos* does not take place in the grain market of Athens.

[52] For this part of the operation, cf. *IG* II^2 1672, lines 292–293, μισθωτοῖς τοῖς ἐκφέρουσι| τὸν σῖτον, τῶν ἑκατὸν μεδίμνων IIII ("hired men who carry out the grain, four obols for 100 *medimnoi*"), i.e., payments to porters.

CHAPTER IV

THE PURPOSE, NATURE, AND IMPLEMENTATION OF THE LAW

OUR KNOWLEDGE of the nature of the $8\frac{1}{3}$% tax rests almost exclusively upon inferences from the contents of Agyrrhios' law of 374/3 B.C. drawn in the light of a variety of comparanda. Since no other ancient source appears to have mentioned this tax, we can only guess about the omission from Agyrrhios' law of such important information as the identity of the people on the islands who are subject to the tax, the method of determining how much wheat and how much barley was taxed and in what proportions, and how the tax was collected. For instance, were farmers required to make a declaration of the size of their harvest or was this somehow established by the tax-collectors?[1] Did local magistrates on the islands participate in verifying the amounts of grain to be taxed and help to collect it, as did the *demarchoi* in the 5th-century B.C. decree on the first-fruits for the goddesses at Eleusis, *IG* I^3 78? Did the farmers have to bring this taxed grain to a central storage facility, possibly near a port, on each of the islands? If the *dodekate* of our inscription had been instituted and described in prior legislation, these and other gaps in Agyrrhios' law might be more explicable.[2]

Perhaps, then, the innovation of our text resides more in the new method of farming the *dodekate* and the *pentekoste*, the handling and storage of the grain once it reaches Athens, and the new provisions for selling it in the Agora to raise revenue for the military fund. It is certainly on these aspects of the taxes that the published record of the *nomothetai* of 374/3 B.C. focused its attention.

At several points in my commentary, I have argued that the *dodekate* on the *sitos* from the islands and the *pentekoste* were produce taxes collected in kind, not in cash. This view provides the most plausible explanation of the opening purpose clause "in order that the *demos* may have *sitos* in the public domain" (lines 5–6), the fact that the *priamenoi* of the tax are excused from the normally required cash down payment (προκαταβολή, line 27), and the definition of the tax in terms of *merides*, each consisting of 400 *medimnoi* of barley and 100 of wheat, and not in terms of the monetary value of the grain.

Since produce taxes collected in kind are not well attested in the Greek *poleis* of the Classical period, little has been known about their implementation.[3] At Athens, however, the procedure at

[1] For the declaration of assets in preparation for levying an *eisphora*, see, e.g., Isokrates 17.49; Demosthenes 27.7; [Demosthenes] 42.18; Harpokration, *s.v.* διάγραμμα; Thomsen 1964, pp. 187–191. Helpful in general on this question with regard to produce taxes is Gofas 1969, pp. 355–364.

[2] Some idea of the complexity raised by these issues in the collection of a produce tax in kind in better documented periods may be gained from the Sicilian tithes (Rickman 1980, pp. 36–42; Nicolet 1994, pp. 215–229). Cf. the fascinating study of Davies (1994) on the Venetian tithes in the Peloponnese. See also briefly, Jones 1974, p. 156. For the possibility that the *dodekate* originated ca. 387/6 B.C., see Salomon 1997, p. 183.

[3] Our law should now serve to qualify views such as "The Romans secured vital grain by taxing their subjects in kind (unlike the Athenians)" (Garnsey 1988, p. 74, cf. pp. 131 and 147); "There is no evidence that many farmers of the city-state had direct, normal tax obligations through assessments either on the value of farmland or on annual revenues from crop production" (Hanson 1995, p. 198); and "in the Greek world quotas on produce were only levied for religious dues" (M. M. Austin, *CAH*^{2} VI, p. 546). Gofas (1969) has shown how it is possible to interpret the 5th century B.C. Καρπολόγοι inscription from Thasos, *IG* XII, Suppl. 349, as evidence for a farmed produce tax collected in kind. For produce taxes collected in kind from Aigai in Aiolis, early 3rd century B.C.(?), see Malay 1983 (*SEG* XXXIII 1034). Another possible example is *SEG* XXXVII 871, from Mylasa. For the produce tithe on crops, wine, and olive oil in the "Zollgesetz der Provinz Asia," see Engelmann and Knibbe 1989, pp. 93–94, lines 72–74. See Gargola 1992 for the produce tax in kind on the grain grown in the Samian Anaia. For produce taxes, see also Gallant 1991, pp. 188–189; Isager and Skydsgaard 1992, pp. 137–141. The whole topic of taxes in kind in the Greek world merits restudy with a full collection of testimonia.

any rate was not unfamiliar, for again the first-fruits decree of the 5th century B.C., *IG* I^3 78, and the accounts of the Eleusinian *epistatai* of 329/8 B.C., *IG* II^2 1672, provide instructive parallels. Although the proportions of wheat and barley taken from individual farmers were small compared to Agyrrhios' $8\frac{1}{3}$%, the first-fruits of the harvest given to Demeter and Kore do in effect constitute a state-imposed produce tax in kind. State officials (*demarchoi* in *IG* I^3 78, some *strategoi* in *IG* II^2 1672) collect the grain. No money changes hands, for the grain is not purchased. The proceeds of this tax in kind are transported to Eleusis to be stored in newly constructed silos (cf. the newly renovated Aiakeion). The grain is measured (and perhaps weighed and tested for quality). Then some of it is used for the *pelanos* and (through barter?) to purchase sacrificial victims. Officials (*hieropoioi*) then sell the residue for cash. In *IG* II^2 1672 the price is set by the *ekklesia*. Like the ten men in Agyrrhios' law, the *hieropoioi* presumably also had to render their accounts and report to the *boule* and/or the Assembly. In the law on the *dodekate*, it is not public officials who collect and transport the proceeds of the tax to Athens. Collection of this particular produce tax was farmed or sold to "purchasers" or tax-farmers who resemble, at least in nomenclature, the men who bought the right to collect such cash-value taxes as the *pentekoste* in the Peiraieus and elsewhere (above, pages 19–20). There is ample evidence from Athens to show that the collectors of the latter tax outbid their competitors at a public auction conducted by the *boule* and the *poletai*. I have inferred from the requirement in line 28 of our law that the *priamenoi* of the $8\frac{1}{3}$% tax pay *kerykeia*, that they too "bought" the right to collect the wheat and barley from the islands in a public auction. Like other tax-farmers, the *priamenoi* of the *dodekate* had to designate guarantors approved by the *boule*. We have seen, however, that this auction is unusual in that the successful bidders were not held responsible for a down payment, either in cash or in kind. Moreover, the sales tax (*eponia*) and the auctioneer's fee (*kerykeia*) were set, again unusually, at a flat rate of twenty drachmai per *meris* and not as a percentage of the cash value of the grain. Indeed the latter was not known at the time of this auction, since the price was to be set by the *demos* no earlier than the beginning of the month Anthesterion, after the grain had been in storage in Athens for approximately three months.

It is striking that, with the exception of this charge of twenty drachmai per *meris*, the law is completely silent about any money changing hands between the *priamenos* and the *polis* at any stage in the procedure. The *priamenoi* do not pay a *prokatabole*. Nothing is said about a schedule of cash payments (καταβολαί) similar to those made by farmers of a cash-value tax. All of the other obligations on the collectors of the *dodekate* are expressed in terms of grain—its quantity, its weight, its quality, the timing of its delivery, and the fact that the risk and expense of its transportation to the Peiraieus and the Aiakeion rest solely with the *priamenoi*.

What then, in this unusual auction, is bid by the prospective collectors of the $8\frac{1}{3}$% produce tax on the grain from the islands and the *pentekoste*? As we have seen, Agyrrhios anticipated that there would be several successful bidders, not just one. If, as seems likely to me, they did not bid in terms of cash and were not held for a down payment in either cash or grain, then it is reasonable to conclude that with their payment of the twenty-drachmai fee for *eponia* and *kerykeia* per *meris* and with the designation of two guarantors, the *priamenoi* of these taxes entered into a contract with the *polis* to deliver a certain quantity of wheat and barley to Athens in clean, dry condition before the month of Maimakterion. In effect, they bid a promise, a promise to carry out the terms of the contract. They were, in a sense, speculating on what we might call grain futures. They were guessing in advance that the wheat and barley harvests on Lemnos, Imbros, and Skyros would be such as to make it worth their investment in the expense of serving as an official collector of the *dokekate* and the *pentekoste* for the *polis*.

I suggest that after the auction, the herald publicly announced the names of the *priamenoi*, the successful bidders whose promises were accepted by the *polis*, and the number of their *merides*, together with the names of their guarantors. The *poletai* probably also recorded this information in writing for their own records, for the *boule*, and eventually for the ten men who were to be elected to supervise the people's grain. This information may also have been distributed to the appropriate officials in the klerouchies of Lemnos, Imbros, and Skyros. It would also not surprise me—though I cannot cite supporting evidence for this guess—if the *priamenoi* of the *dodekate* and the *pentekoste* also received from the *polis* some kind of written authorization, perhaps marked by the public seal, identifying them as official collectors of the produce taxes in the current archonship. Something of the sort would seem to have been necessary to distinguish the *priamenoi* of the *dodekate* and the *pentekoste* from other *emporoi* involved in purchasing grain on the islands.

In the commentary I have tried to show that the lawgiver cast a long section of his law in the form of a contract between the *polis* and the *priamenoi* of these taxes (lines 5–36). The advantages reaped by the former from successful execution of the contract are obvious: the *demos* is to have grain in the public domain which will be deposited in the Aiakeion at the risk and expense of the tax-collectors before the month of Maimakterion. Within thirty days of its arrival at this facility, the grain will be weighed, measured, and tested for quality under the supervision of publicly elected officials according to specifications set by the *polis*. Earlier, at the time of the auction, the *polis* will collect *eponia* and *kerykeia* from the successful *priamenoi* and protect itself against potential default in the execution of the contract by registering two guarantors per *meris* of grain. In the case of a *symmoria*, the *polis* explicitly establishes its right to exact grain from all six members individually and as a group. After the *priamenoi* have fulfilled their contract obligations, the *polis* continues to benefit through the sale of the people's grain and the revenue thereby produced, which is allocated to the military fund.

In return for having its tax-grain deposited in the Aiakeion in the requisite quantity and quality, the *polis* undertakes to provide this new public granary, protected by a roof and door(s), to the *priamenoi* rent-free for a maximum period of thirty days during which they are to weigh out their grain. The *polis* excuses the *priamenoi* from a down payment when they "purchase" the taxes. Beyond this, the *polis* seems to have had few other obligations to the *priamenoi*, except perhaps, as I have suggested, to provide them and/or the farmers on the islands with some form of official recognition as state-approved collectors of the *dodekate*.

On the other side of the contract, it is the *priamenoi* who assume all of the risks of losses at sea, or other dangers of theft and spoilage, and so forth, and the costs of collecting the taxes on the islands and transporting the grain to the Peiraieus and up to the city. They had to meet a deadline for delivery of the grain and for completing the weighing, measuring, and quality control within thirty days after bringing it up to Athens. At the time of the "selling" of the *dodekate* and *pentekoste* taxes by the *polis*, these men had to pay a fixed fee of twenty drachmai and provide two guarantors per *meris*. In addition to these, and possibly other heavy obligations spelled out in other legislation, such as the τελωνικοὶ νόμοι (above, pages 28–29), the *priamenoi* at the time of the auction had to be familiar enough with the potential productivity of one or more of the islands of Lemnos, Imbros, and Skyros in both wheat and barley, and with local climatic conditions in a given year to make intelligent bids as to the number of *merides* they would contract to deliver to Athens. I have argued above, pages 60–61, that for this and other reasons it is probably a safe inference that many or most of the tax-collectors were grain merchants, some of them possibly non-Athenian. For an *emporos* resident in Skyros in the 4th century B.C. with close contacts to the Athenian banking community, see [Demosthenes] 52.3.

Given the nature of most forms of taxation at Athens and in other Greek *poleis* of the Classical period, men of the type just described were essential to the success of Agyrrhios' law, since the *polis* was not equipped to collect its own taxes. Athenian taxes, as opposed to tribute, were almost always collected not by state officials, but by tax-farmers. Although it provided the Aiakeion rent-free, waived a down payment, and even called for the election of a new board of public officials to supervise its tax-grain once it reached Athens, the *demos* in the normal manner entered into a contract with private individuals for the collection of the *dodekate* and the *pentekoste*.

Agyrrhios and the Athenian *nomothetai* were seeking to appeal to men characterized by Xenophon, *Oikonomikos* 20.27–28, as follows: οἱ ἔμποροι φιλόσιτοί εἰσι. καὶ γὰρ οἱ ἔμποροι διὰ τὸ σφόδρα φιλεῖν τὸν σῖτον, ὅπου ἂν ἀκούσωσι πλεῖστον εἶναι, ἐκεῖσε πλέουσιν ἐπ' αὐτὸν καὶ Αἰγαῖον καὶ Εὔξεινον καὶ Σικελικὸν πόντον περῶντες· ἔπειτα δὲ λαβόντες ὁπόσον δύνανται πλεῖστον ἄγουσιν διὰ τῆς θαλάττης, καὶ ταῦτα εἰς τὸ πλοῖον ἐνθέμενοι ἐν ᾧπερ αὐτοὶ πλέουσι. καὶ ὅταν δεηθῶσιν ἀργυρίου, οὐκ εἰκῇ αὐτὸν ὅπου ἂν τύχωσιν ἀπέβαλον, ἀλλ' ὅπου ἂν ἀκούσωσι τιμᾶσθαί τε μάλιστα τὸν σῖτον καὶ περὶ πλείστου αὐτὸν ποιῶνται οἱ ἄνθρωποι, τούτοις αὐτὸν ἄγοντες παραδιδόασι ("Merchants are lovers of grain. To be sure, because their love of grain is so intense, the merchants sail after it to whatever place they hear has a surplus, crossing the Aigaion, the Euxine, or the Sicilian Sea. Then after they get their hands on as much as they possibly can, they transport it over the sea, loading it into the same ship on which they are themselves sailing. And whenever they want money, they don't randomly throw the grain away wherever they happen to be, but wherever they hear that grain is fetching the highest price and where people are most desperate for it, that's where they carry it and make it available").

Kevin Clinton has suggested to me that each *priamenos* may have been assigned a specific area in one of the islands in which he could collect the tax. To discourage *priamenoi* from wandering all over Lemnos, Imbros, and Skyros at will and collecting wheat and barley as fast as they could, perhaps the Athenians had established zones of taxation on the islands. Bidding in the auction in the *boule* might have been conducted with these designated zones kept clearly in mind. Or perhaps successful bidders were assigned certain zones by the *poletai*, though this seems to me unlikely. At any rate, Agyrrhios' law as inscribed on our stele offers no information on this potentially important point.

One of the most puzzling aspects of our new law, which will also, I predict, generate much future discussion and controversy, is how collection of the *dodekate* and the *pentekoste* was made attractive enough to encourage participation in the auction from competitive bidders numerous enough to account for all the *merides* of grain which the *polis* expected to realize from these taxes in the current year. For *emporoi*, surely one of the most effective ways to bring this about would have been to ensure potential bidders a profit in collecting the tax and transporting portions of the people's grain to Athens. There may have been other inducements as well, for all we know, but profit is likely to have remained the principal motive for grain merchants to participate in the farming out of these taxes.

Not a word, however, is said in Agyrrhios' law about the profits of the *priamenoi*. This should not surprise us, given the nature of most Athenian taxation. In formulating regulations on taxation, the *polis* was primarily concerned with getting its estimated revenue and protecting its citizens from extortion. The tax-farmer's profits were his own concern and almost never figure explicitly in Athenian tax legislation.[4] For almost all the taxes that were farmed out or sold at auction, profits resulted from the tax-collector's informed speculation on the volume of trade. This the *polis*, for obvious reasons, almost never sought to restrict.

The fact that tax legislation from ancient Greece is normally silent about the tax-farmer's profits complicates the task of historians seeking to reconstruct the implementation of such taxes. It is doubly difficult to do so with produce taxes in kind, which, outside Egypt, are very poorly documented in our surviving sources. Trying to infer how the *priamenoi* of the *dodekate* made a profit is even more hazardous because, as we have seen, this tax is not mentioned elsewhere. We are reduced once more to drawing inferences from the wording and lacunae of the text and to employing helpful comparanda.

[4] For similar "gaps" in Athenian legislation governing maritime commerce, see the discussions conveniently cited by Gofas (1993, pp. 233–234).

One possible source of income or profit for the *priamenoi* of the *dodekate* and the *pentekoste* might have been a cash payment from the *polis* for their services. The text of the law itself, however, lends no support to such a conjecture. Although it registers payments of *eponia* and *kerykeia* from the *priamenos* and explicitly states that transportation of the grain from the islands to Athens is to be κινδύνωι τῶι ἑαυτõ ... καὶ ... τέλεσιν τοῖς α̣[ὐ]τõ, there is no indication in the law that any other money changed hands between the *priamenos* and the *polis*.

In seeking a plausible occasion for profit by the *emporoi/priamenoi*, and possibly even payment by the *polis*, it might be best to begin with the period after the grain reaches Athens, for it is regarding procedure during this time that the law is most informative. It seems very unlikely that at the end of what we might call one tax-cycle, the *priamenoi* received a payment out of the proceeds of the public sale of the wheat and barley by the ten newly elected magistrates in the Agora. The elaborate instructions in lines 51–55 of the law regulating this sale omit all mention of such a payment to the tax-collectors at this stage in the process. In fact, when the ten men carry their bags of coins up to the Pnyx, all the proceeds of the sale, τὰ ἐκ τõ σίτο γενόμενα (line 55), are specifically designated for the military fund, not merely what is left after presumed—but unstated—fees for the tax-collectors had been deducted. It seems, then, that the opportunity or occasion for payment to or profit by the tax-farmers must have arisen at some earlier stage in the procedure.

A suitable earlier occasion for profit might have occurred when the ten newly elected magistrates had completed their weighing out and inspection of the grain in the Aiakeion. They presumably at that time verified that each tax-farmer had in fact delivered his portion(s) of wheat and barley on time and in the required condition. We are not told explicitly that such verification by the ten men marked the end of each tax-farmer's contract obligations. It is possible that first the ten men had to report their findings to the Council of Five Hundred and to the *poletai*, who then verified the accounts and officially announced that the tax-farmers had fully discharged the terms of their contract with the *polis*. No hint of such a requirement survives in the text of our law, but this may have been part of normal operating procedure that was covered in other prior regulations. It is even conceivable, I suppose, that instructions for payment to the tax-collectors for delivering the people's grain from the islands to Athens were also contained in other previous regulations. Total omission of such an important item, however, without even as much as a brief cross-reference in our law to existing regulations, seems to me highly unlikely.

The effect of this speculation is to push back the occasion for the tax-farmers' profit to an earlier stage in the process. In my view, the most likely occasion would have been before, rather than after, the grain was transported to Athens. If we can invoke an analogy from tax-farming which involves the collection of cash, we see that it is in the collection process itself that profits can be made. The farmer of a tax assessed in cash speculates on the *volume* of trade in a given year. He is not permitted in any individual transaction to extract more than the official rate of taxation set by the state, but he estimates that there will be enough transactions from which he can extract that official rate to produce a sum of money in excess of the figure he had guaranteed to pay to the state. Herein lies his profit. The state is interested primarily in receiving the sum that the highest bidder in the auction has contracted to pay to it. As long as he does not break the law by extorting more than the official rate from any one taxpayer, as long as he meets his required schedule of payments, his profit margin seems to have remained outside of the concern of the *polis* and free from state regulations.

Applying this analogy to the grain-tax law, we need to ask what the *priamenos* did when he sailed to Lemnos, Imbros, and Skyros after being recognized as an official tax-collector of the *polis*. Here the law leaves us completely in the dark. Among the many questions we have not answered is whether he had to negotiate *in locis* with individual farmers or whether the tax-grain was first brought into a central depository, perhaps under the supervision of local officials before negotiations began. In the midst of all these unknowns,[5] we can only guess. But let us suppose, for example, that a *priamenos* has contracted to deliver to the Aiakeion in Athens at the specified time one portion of 500 *medimnoi* of grain, which will occupy roughly one-sixth of the capacity of a merchant ship. If this is a true produce tax paid in kind, in which the *polis* is taking $8\frac{1}{3}\%$ of the harvest from Athenian farmers on the islands, our *priamenos* will extract this percentage from one or more farmers until he fills up the portion of 500 *medimnoi* he owes the *polis*. Up to this point, of course, he has made no profit. But if he was a grain merchant, he may have been in an excellent position to negotiate the sale of some of the remaining $91\frac{2}{3}\%$ of a particular farmer's harvest at a price favorable to himself. He also may have been able to extract more grain, still at the official rate of $8\frac{1}{3}\%$, from other farmers over and above the portion that he has already filled. For all we know, he may have been able to keep some of this tax-grain as his own and sell it for his profit.

All this is pure guesswork, of course, and there may have been all kinds of other complex deals struck by the tax-collectors with grain producers on the islands. All that we know from the law is that the *polis* set a fixed rate of taxation. Tax-collectors could not extract more than this percentage from any one transaction without facing probable prosecution. Also, by "selling" a certain number of *merides* to the *priamenoi*, the *polis* knew exactly how much grain it was going to receive. It too may have fixed the number of *merides* on the basis of its own estimation of the yield of grain in the three islands in a given year. If the yield turned out to be higher, the tax-collectors probably got to keep the surplus. After having already arrived at its predicted total of *merides* and completing the auction, the *polis* was probably not in a position at a later stage to demand more *merides* from the *priamenoi*.

If, on the other hand, the harvest on the islands in a given year failed to produce enough tax-grain to match the number of *merides* auctioned off in Athens, it seems likely that some *priamenoi* would have fallen short of the quota they had contracted to deliver to the Aiakeion. In this case we may guess that they had to make up the shortfall with wheat and barley that they purchased from grain producers on the islands. It is difficult to see how otherwise they could have fulfilled their contractual obligations with the *polis*.

Admittedly, almost all of my tentative reconstruction of this situation is inferential. The law does not spell it out. I would suggest that the silence on these points is intentional. In his law Agyrrhios here transformed a tax on the produce of Lemnos, Imbros, and Skyros into a highly speculative scheme designed to give the *demos* its own grain, to bolster the military fund with additional income, and to attract grain merchants and businessmen eager for potential new profits, into working for the *polis*. The latter, while making specific and not too onerous demands on its contractual partners, also freely entered into speculation about its projected income derived from grain futures. But the law explicitly lays down the conditions whereby the *polis* is protected from loss.

The ground rules in the law for the *priamenoi*, on the other hand, concentrate almost exclusively on their minimum obligations to the *polis*. Participation in the scheme entailed for them expenses, risks, and possibly fines or other penalties for default on their contract. Like the *polis*, they had to guess about grain futures in the islands, but they were also left free to read between the lines of this bold new legislation to discover ways in which service as an official tax-collector of the *polis* could be turned into increased profits. At a minimum, authorization by the *boule* to collect $8\frac{1}{3}\%$ of their harvest forced farmers to deal with the *priamenoi* of the *dodekate*. This might in many cases give the latter a competitive edge over other private *emporoi*. Agyrrhios probably also foresaw many other advantages to be gained by imaginative and daring *priamenoi*, but he was too shrewd a businessman himself and too skilled a politician to make any of this explicit in his law.

[5] Among these are also whether the *emporoi/priamenoi* in every instance sailed personally to the islands or whether they were represented by agents. See my speculation above about the members of a *symmoria*, pp. 65–66. Individual *priamenoi* might even have specialized in one of the islands, perhaps because they had detailed local knowledge or contacts with friends, relatives, or business associates there.

There was another possible source of profit for the collectors of the grain-tax from the islands: they almost certainly did not sail out to Lemnos, Imbros, and Skyros in empty ships but instead probably found room for cargo from Athens that could be sold at a profit at their final destination in the three islands or along the way. The *priamenoi* might also have tapped into what was probably a lively passenger trade between the mother city and her three island dependencies.

Until more evidence is available, details about profits and other possible incentives for *emporoi* to participate in the farming of the *dodekate* and the *pentekoste* from the islands must remain speculative. We also have little or no grounds for assessing another important issue raised by this inscription: did the law work? Were the elaborate procedures for farming the *dodekate* and the *pentekoste*, for renovating the Aiakeion, for providing the *demos* with grain in the public domain, and for selling the people's grain to supplement the military fund effective and for how long? Did this *dodekate* remain an essential element in the grain supply and tax structure of 4th-century B.C. Athens, or could the troubling and apparently total silence of our surviving sources suggest that Agyrrhios' law did not succeed? Even if it was annulled or simply ignored, its survival on stone has nevertheless brought us a great deal of new evidence about Athenian lawmaking and the economy which will be of permanent value to scholars.

The tenses of the verbs in this law, repeatedly in the future indicative, or the present infinitive and imperative, together with frequent present general conditional clauses in the subjunctive mood, indicate that the Athenians looked forward to many years of collecting a produce tax in wheat and barley from the three islands. There is no compelling reason to view this law as a response to a specific emergency or to regard one of the islands as singled out for special treatment.

Finally, we cite what is perhaps one of this law's most important contributions to our understanding of the economy of Classical Athens and possibly other *poleis:* By selling the *dodekate* and the *pentekoste* to private individuals, the Athenians assured that every merchant ship carrying as little as one *meris* of the people's wheat and barley had to sail to the Peiraieus. Surely, therefore, an unstated consequence of such a contract was that *priamenoi* were thereby encouraged to fill up the rest of their ships with grain bound for Athens. The *polis* in effect not only legislated for the guaranteed delivery of its own tax-grain but also took indirect steps to increase the overall volume of wheat and barley that could be shipped to the Peiraieus. For all we know, some of this grain might otherwise have gone to other destinations. After turning his *meris* over to the ten men, a *priamenos* was free to sell the rest of his cargo to the grain dealers on the open Athenian market.

One is prompted at this point to ask whether in envisaging this kind of transaction Agyrrhios' entrepreneurial experience led him to establish the size of each *meris* of tax-*sitos* at roughly one-sixth the capacity of a merchant ship. This would in each case leave the *priamenos* free to make a profit on the rest of his cargo, particularly if his privileged status as an official tax-collector had in fact enabled him to buy this grain at a favorable rate on the islands. Could we even carry this speculation a step further and suggest that in keeping the size of a *meris* at this modest level, the *polis* not only maximized the opportunity for profit by the *priamenoi* on the sale of the rest of their cargo at Athens but also intended to increase the number of potential *priamenoi*? That is, if we are correct in concluding that *priamenoi* were not paid a fee for collecting the *dodekate* and that there seems to have been little opportunity for profit in the process of extracting 500 *medimnoi* of tax-*sitos* from one or more farmers on the islands, then there might have been little incentive for tax-farmers to bid for more than one *meris*. The *nomothetai* did not include in this law any restriction on the number of *merides* for which each *priamenos* could bid, but the *polis* might not have been distressed if many of them contracted for only one. The more nontax-grain in his ship, the more potential profit for the *emporos*. The more *merides* the *polis* auctioned off one at a time, the more *priamenoi* who would participate in the tax-collection process and, of course, the more grain ships that would of necessity be sailing to the Peiraieus. At the cost of a modest fee of twenty drachmai for *eponia* and *kerykeia* and by naming two guarantors, a *priamenos* secured privileged status in his negotiations with grain producers in the islands. Although additional ways to profit are not spelled out in the new regulations, experienced *emporoi* would need little coaching to devise them while at the same time gaining public recognition in Athens as servants of the *polis*.

All of the foregoing is, of course, unsupported conjecture. I am inferring unstated intentions in the new law on the basis of imagined consequences. But I do not believe that the proposed reconstruction goes beyond the ingenuity of either the lawgiver or the businessmen he wanted to enroll in his new scheme. Like many successful politicians, Agyrrhios appears to have aimed at providing something for everyone: the *demos* will have its own public grain at a price it will itself determine; the military fund will gain increased revenue from this new source; more grain ships will be attracted to the Peiraieus; members of the "world of the emporion" will make more money while serving the *polis*.

The broader implications of this reconstruction bring, I think, a different perspective to the long-vexing problem of the degree of intervention of the *polis* in overseas commerce in the importation of grain. The Athenian *ekklesia* certainly legislated on such topics as the necessity for Athenian grain merchants to bring their cargoes to Athens and the requirement that they bring two-thirds of each cargo up to the *asty* (see Gauthier 1981, pages 19–28). The new Agora inscription, too, contains evidence of direct legislative control over the means of collecting and transporting to Athens significant quantities of grain. This was still, of course, in a sense tax-grain that belonged to the Athenian *demos*, and we should not be surprised to find explicit published regulations governing the agents who entered into a contract with the *polis* concerning the *dodekate* and the *pentekoste*. But Agyrrhios and the *polis* also seem clearly to be working in more subtle and indirect ways in this law to exert influence over trade in wheat and barley from the islands, to encourage merchants and shipowners to bring more and more grain to Athens.

Finally, it is appropriate to try to place this new law in the development of the little we know about the role of the *polis* in the importation and sale of grain. To my knowledge, lines 5–6 are the earliest surviving evidence at Athens for direct action by the *polis* to provide its citizens with a regular annual supply of wheat and barley. It is possible that Agyrrhios introduced a new concept into the Athenian economy—a constant supply of public grain (δημόσιος σῖτος) complementing (ὁ ἐν ἀγορᾷ σῖτος) the private grain of the market. To be sure, the quantity of the tax-grain transported annually to Athens from Lemnos, Imbros, and Skyros filled only a small percentage of the city's yearly needs. But the yield from these two taxes, as we have seen, was not insignificant[6] and its public sale in the critical season after Anthesterion probably had an impact on the price of grain in the market. Thus price-fixing by the *polis* represented a significant intervention into what was normally regarded as the private sector.

Not only do the provisions of the grain-tax law seem remarkably innovative, from our uninformed vantage point, but there is also a considerable interval in time and in our sources before we learn of later attempts by the *polis* to help provide Athenians with grain. To my knowledge these do not emerge in our testimonia until ca. 357 B.C., when apparently Kallisthenes was σιτώνης (Demosthenes 20.33–34), an office to which Demosthenes was elected after 338 B.C. (Demosthenes 18.248; Deinarchos 1.80). We know too little about the office at this time to conclude anything about regularity of election to it and the duties of the incumbent, although Demosthenes clearly seems to have been provided with eight talents and sent abroad to buy grain for the *demos*.[7]

Recent research by Migeotte, Strubbe, and others[8] has shown that *ad hoc* appointment of *sitonai* in times of crisis in the food supply gradually developed in many cities of Hellenistic Greece into regular magistracies accompanied by annual allocations from the public "budget" for the purchase of public grain. One of the earliest known *poleis* to have adopted these practices was Teos

[6] For a rough estimate, see above, p. 76.

[7] For *sitonai* at Athens, see Fantasia 1987; Couilloud-La Dinahet 1988.

[8] See above, p. 68, note 159.

ca. 300 B.C.[9] As far as we can tell, evidence for this kind of machinery at Athens seems not to predate 283/2 B.C., *IG* II^2 1272, lines 3–4; cf. *IG* II^2 792.

It is instructive, therefore, to set the very detailed and elaborate provisions of Agyrrhios' law of 374/3 B.C., together with its clearly stated purpose in lines 5–6, ὅπως ἂν τῶι δήμωι σῖ[το]|ς ἧι ἐν τῶι κοινῶι, into this historical setting. It is a fair inference that the city that probably imported more grain each year than any other Greek *polis* (Demosthenes 20.31) may have developed significant innovations for exercising some control over its own vital food supply much earlier than most of us would have assumed. Moreover, we must not forget that it is to the recent discovery of a single new inscription that we owe our knowledge of the relatively early date at which these maneuvers are now attested and the fact that we must revise our whole conception of grain-taxes in kind providing food for the *demos*.

9 *Syll.*^{3} 344; Welles 1934, pp. 15–23, no. 3, lines 72–94; Bresson 1993, pp. 168–169. For possible regulation of the grain trade in Teos in the 5th century B.C., see Meiggs and Lewis, *GHI*, no. 30; additional fragments in *SEG* XXXI 984 and 985; the latter do not concern *sitos*.

CHAPTER V

THE HISTORICAL SETTING OF THE LAW

IT IS TIME NOW to consider the historical setting of this important new document. Amid other perennial reminders, the Athenians in late summer of 376 B.C. encountered a blunt and dangerous demonstration of the precariousness of their imported supply of grain. A large number of grain transports heading for the Peiraieus had reached Geraistos on the southern tip of Euboia when a Spartan fleet of sixty triremes under the command of the *nauarchos* Pollis established bases at Andros, Keos, and Aigina and sailed into the Saronic Gulf. Reluctant to risk sailing past this Spartan naval blockade, the grain ships hung back at Euboia; Athens was in effect under siege. It was not until Chabrias led out an Athenian fleet that the blockade was broken and the Spartans were decisively defeated in a naval battle at Naxos, clearing the way for the grain ships to sail into their original destination.[1]

Despite this inspiring victory and another the following year won by Timotheos over a large Spartan fleet near Leukas, the strain of financing their naval operations was taking its toll on the Athenians.[2] Piracy in the Saronic Gulf continued to harass the grain ships bound for the Peiraieus which were so vital to the Athenian economy. In an often-quoted passage (*HG* 6.2.1), Xenophon briefly assesses Athens' dilemma: οἱ δὲ Ἀθηναῖοι, αὐξανομένους μὲν ὁρῶντες διὰ σφᾶς τοὺς Θηβαίους, χρήματά τε οὐ συμβαλλομένους εἰς τὸ ναυτικόν, αὐτοὶ δὲ ἀποκναιόμενοι καὶ χρημάτων εἰσφοραῖς καὶ λῃστείαις ἐξ Αἰγίνης καὶ φυλακαῖς τῆς χώρας, ἐπεθύμησαν παύσασθαι τοῦ πολέμου, καὶ πέμψαντες πρέσβεις εἰς Λακεδαίμονα εἰρήνην ἐποιήσαντο ("The Athenians, seeing that the Thebans' power was increasing through their efforts and that they were not contributing money for the fleet, while they themselves were being worn down by *eisphorai* taxes and piratic raids from Aigina and by garrisoning their territory, were eager to put an end to the war, so after sending envoys to Lakedaimon, they made peace"). This so-called Peace of 375/4 B.C. was of very short duration. Scholars have wrangled over its exact date and the confusing sequence of events in the immediately subsequent years, especially relations with Thebes and the exploits and travails of the general Timotheos. For our purposes here it is enough to note that burdensome taxes, endangered shipments of grain, and a shrinking military fund were all probably frequent topics in debates in the *ekklesia* at this time. In this kind of military and economic climate, the Athenians appointed the *nomothetai* who brought in Agyrrhios' law on the *dodekate* and the *pentekoste* on the grain from the islands.[3] At the same time, after the Peace, the Athenian *demos* might have been in a slightly more buoyant mood. Munn (1993, pages 172–183, 229–230, citing Isokrates 15.109–110, 4.20) views the Peace as "cause for joy and thanksgiving on the part of the Athenians," who saw in the respite it brought from hostilities against the Spartans a "rebirth of Athenian ascendancy."

Its minimal dating formula ἐπὶ Σωκρατίδο ἄρχοντος in line 2 does not permit more precise placement of this law within the year 374/3 B.C. Although the text itself contains a number of time designations, as we have seen (lines 17, 38–39, 43–44, 48, 50–51, 58–60), they do not appear to help us narrow down the interval during which the law was passed. Helpful adverbs such as

[1] Xenophon, *HG* 5.4.60–61; Demosthenes 20.77; 22.15(?); Diodoros 15.34.3–35.2; Plutarch, *Phokion* 6; *Camillus* 19.6; Polyainos 3.11.2. For an analysis of the interplay between the accounts of Xenophon and Diodoros on these events, see Tuplin 1993, p. 159; for the finances, Brun 1983, pp. 39–48.

[2] ὁ δὲ Τιμόθεος ... χρήματα μέντοι μετεπέμπετο Ἀθήνηθεν· πολλῶν γὰρ ἐδεῖτο, ἅτε πολλὰς ναῦς ἔχων ("Timotheos now sent for cash to be brought from Athens, for he needed a great deal inasmuch as he possessed many ships"), Xenophon, *HG* 5.4.66.

[3] For the general situation, see Harding 1988. For discussion of the date, see Cawkwell 1963b; Buckler 1971; Sealey 1993, p. 285, note 59.

αὐτίκα μάλα and the like, which might have indicated urgency, are lacking. Nor do we know enough about the procedure for appointing *nomothetai* at this period to fix a *terminus post quem* during this archonship for their activities. If Agyrrhios aimed at implementation of the auction of the *dodekate* and the *pentekoste*, renovation of the Aiakeion, election of the ten men, and all the concomitant activity outlined on the stele still within the archonship of Sokratides, then the law must have been passed in the first months of the year. Otherwise, the interval before the deadline for delivery of the grain before Maimakterion, the fifth month in the year, would probably not have left enough time for the auction and the alterations to the Aiakeion. Also, the election of the *strategoi* for 374/3 B.C. had probably already taken place in the spring of the previous archon-year, and there would not be another occasion concurrently to elect the first board of ten men to supervise the grain until the spring of 373 B.C. It seems more probable then that the sequence Agyrrhios had in mind was something like (1) election of the ten men after the sixth prytany of the archonship of Sokratides, spring 373 B.C., to take office at the beginning of the next archon-year, Asteios, in 373/2 B.C.; (2) auction of the *dodekate* and the *pentekoste* probably in the spring of 373 B.C., before the harvest; and (3) deadline for delivery of the *merides* of grain from the first cycle of collecting the taxes in kind before Maimakterion of the archonship of Asteios in autumn of 373 B.C. Work on the alterations to the Aiakeion could have begun as soon as the law was carried; the new granary had to be ready by summer of 373 B.C. The wheat and barley from Lemnos, Imbros, and Skyros remained in storage from Maimakterion to Anthesterion (at the earliest), over the winter of 373/2 B.C., the archonship of Asteios. The ten men would then have conducted the first public sale of the people's grain in the late winter/early spring of 372 B.C. We have suggested above, pages 80–81, that in lines 55–61 Agyrrhios made provision for a year of transition during which the *dodekate* and *pentekoste* taxes would continue to be collected in cash.

Nothing is said in the law about how many days the ten men were to spend in selling the people's grain. It is not possible to determine, therefore, exactly when they deposited the cash proceeds of the first sale with the assembly, although it could still have been within the month of Anthesterion, 372 B.C. Allocation of this money to the stratiotic fund could have been swiftly effected by the *apodektai*, so that Athenian *strategoi* engaged in the precariously funded work of the young "Second Athenian Alliance" might have used it almost immediately. We have suggested above, page 76, that one result of the new law was that roughly more than eighteen talents per year became available. We should not regard this sum as a windfall, but as additional income, ingeniously planned for, in a period when the contributions of Athens' allies may not have been entirely reliable.

* * *

I publish this inscription knowing that I have left unresolved many of the important issues it raises. It is also clear that one could spend several more years exploring and commenting upon the rich and manifold contributions this remarkable document makes to our understanding of the ancient world. For instance, no attempt has been made here to examine how the law of Agyrrhios might have anticipated later regulations for tax-farming and produce-taxes in Ptolemaic Egypt and Sicily under the Romans. I now leave these pleasures to others.

REFERENCES

Adams, S. 1989. "Aspects de la sécurité de la navigation dans l'antiquité grecque," *Symposion 1985*, Cologne, pp. 283–291.

Adler, A. 1928–1938. *Suidae Lexicon*, 5 vols., Leipzig.

Agora = *The Athenian Agora: Results of Excavations Conducted by the American School of Classical Studies at Athens*, Princeton

III = R. E. Wycherley, *Literary and Epigraphical Testimonia*, 1957

X = M. Lang and M. Crosby, *Weights, Measures, and Tokens*, 1964

XIV = H. A. Thompson and R. E. Wycherley, *The Agora of Athens: The History, Shape, and Uses of an Ancient City Center*, 1972

XVI = A. G. Woodhead, *The Inscriptions: The Decrees*, 1997

XIX = G. V. Lalonde, M. K. Langdon, and M. B. Walbank, *Inscriptions: Horoi, Poletai Records, and Leases of Public Lands*, 1991

XXI = M. Lang, *Graffiti and Dipinti*, 1976

XXVI = J. H. Kroll, *The Greek Coins*, 1993

XXVII = R. F. Townsend, *The East Side of the Agora: The Remains beneath the Stoa of Attalos*, 1995

XXVIII = A. L. Boegehold et al., *The Lawcourts at Athens: Sites, Buildings, Equipment, Procedures, and Testimonia*, 1995

Alessandrì, S. 1982. "Alcune osservazioni sui segretari ateniesi nel IV sec. a.C.," *AnnPisa* 12, pp. 7–70.

Allen, R. E. 1971. "Attalos I and Aigina," *BSA* 66, pp. 1–12.

Amouretti, M. C. 1979. "Les céréales dans l'antiquité: Espèces, mouture et conservation, liaison et interférences dans la Grèce classique," in *Les techniques de conservation des grains à long term*, M. Gast and F. Sigaut, eds., Paris, pp. 57–69.

———. 1986. *Le pain et l'huile dans la Grèce antique*, Paris.

Ampolo, C. 1986. "Il pane quotidiano delle città antiche fra economia e anthropologia," *Opus* 5, pp. 143–151.

Amyx, D. A. 1958. "The Attic Stelai, Part III: Vases and Other Containers," *Hesperia* 27, pp. 163–307.

Andreades, A. M. 1933. *A History of Greek Public Finance*, I, C. N. Brown, trans., Cambridge, Mass.

Andrewes, A. 1937. "Athens and Aegina, 510–480 B.C.," *BSA* 37, pp. 1–7.

Antoniades, C. A. 1977. Η Σκύρος στους περιηγητές και γεωγράφους (1400–1900) [Παράρτημα Αρχείον Ευβοικών Μελετών 21].

Battaglia, E. 1989. *'Artos': Il lessico della panificazione nei papyri greci* [*Biblioteca di Aevum Antiquum* 2], Milan.

Behrend, D. 1970. *Attische Pachturkunden* [*Vestigia* 12], Munich.

Beloch, K. J. 1912–1927. *Griechische Geschichte*, 2nd ed., I–II, Strassbourg, 1912–1916; III–IV, Berlin and Leipzig, 1922–1927.

Billot, M.-F. 1992. "Le Cynosarges, Antiochos, et les tanneurs: Questions de topographie," *BCH* 116, pp. 119–156.

Bloch, A. 1975. "Literarische und inschriftliche Gesetzprosa in Griechischen," *MusHelv* 32, pp. 135–154.

Boeckh, A. 1886. *Die Staatshaushaltung der Athener*, 3rd ed., I–II, rev. M. Fränkel, Berlin.

Boegehold, A. L. 1967. "Philokleon's Court," *Hesperia* 36, pp. 111–120.

Boerner, J. *RE* VI.i, 1907, col. 243 ('Επώνιον).

Bogaert, R. 1968. *Banques et banquiers dans les cités grecques*, Leiden.

Bongenaar, J. C. A. M. 1933. *Isocrates' Trapeziticus*, Utrecht.

Bousquet, J. 1989. *Corpus des inscriptions de Delphes* II: *Les comptes du quatrième et du troisième siècle*, Paris.

Bresson, A. 1993. "Les cités grecques et leurs *emporia*," in *L'Emporion*, A. Bresson and P. Rouillard, eds., Paris, pp. 163–226.

Brun, P. 1983. *Eisphora-Syntaxis, Stratiotika: Recherches sur les finances militaires d'Athènes au IVe siècle av. J. C.* [*Annales littéraires de l'Université Besançon* 284, *Centre de Recherches d'histoire ancienne* 50], Paris.

Buchanan, J. J. 1962. *Theorika*, Locust Valley, N.J.

Buckler, J. 1971. "Dating the Peace of 375/4 B.C.," *GRBS* 12, pp. 353–361.
Bugh, G. R. 1988. *The Horsemen of Athens*, Princeton.
Bullock, C. J. 1939. *Politics, Finance, and Consequences* [*Harvard Economic Studies* 65], Cambridge, Mass.
Burford, A. 1993. *Land and Labor in the Greek World*, Baltimore.
Burke, E. M. 1992. "The Economy of Athens in the Classical Era," *TAPA* 122, pp. 199–226.
Busolt, G. 1893–1904. *Griechische Geschichte*, 2nd ed., I–III, Gotha.
Camp, J. McK. II. 1979. "A Drought in the Late Eighth Century B.C.," *Hesperia* 48, pp. 397–411.
———. 1982. "Drought and Famine in the 4th Century B.C." in *Hesperia* Supplement 20: *Studies in Athenian Architecture, Sculpture, and Topography Presented to Homer A. Thompson*, Princeton, pp. 9–17.
———. 1986. *The Athenian Agora: Excavations in the Heart of Classical Athens*, London.
———. 1990. *The Athenian Agora: A Guide to the Excavation and Museum*, 4th ed., Athens.
Cargill, J. 1995. *Athenian Settlements of the Fourth Century B.C.* [*Mnemosyne* Supplement 145], Leiden.
Carnes, J. S. 1995. "Why Should I Mention Aiakos? Myth and Politics in Pindar's *Nemean* 8 [Part I]," *QuadUrb* 51, pp. 7–48.
Cartledge, P., P. Millett, and S. Todd. 1990. *Nomos: Essays in Athenian Law, Politics, and Society*, Cambridge.
Casson, L. 1971. *Ships and Seamanship in the Ancient World*, Princeton.
———. 1991. *The Ancient Mariners*, 2nd ed., Princeton.
Cawkwell, G. L. 1962. "Demosthenes and the Stratiotic Fund," *Mnemosyne* 15, pp. 377–383.
———. 1963a. "Eubulus," *JHS* 83, pp. 47–67.
———. 1963b. "Notes on the Peace of 375/4," *Historia* 12, pp. 84–95.
———. 1975. "Thucydides' Judgment on Periclean Strategy," *YCS* 24, pp. 53–70.
Chantraine, P. 1940. "Conjugaison et histoire des verbes signifiant vendre (πέρνημι, πωλέω, ἀποδίδωμι, ἐμπολῶ)," *RPhil* ser. 3, 14, pp. 11–24.
Charneux, P. 1984. "Phratries et Kômai d'Argos," *BCH* 108, pp. 207–227.
Clinton, K. 1971. "Inscriptions from Eleusis," Ἀρχ Ἐφ, pp. 81–136.
———. 1994. "The Eleusinian Mysteries and Panhellenism in Democratic Athens," in *The Archaeology of Athens and Attica under the Democracy*, W. D. E. Coulson, O. Palagia, T. L. Shear Jr., H. A. Shapiro, and F. J. Frost, eds., Oxford, pp. 161–172.
Cohen, E. E. 1989. "Athenian Finance: Maritime and Landed Yields," *ClAnt* 8, pp. 207–223.
———. 1992. *Athenian Economy and Society: A Banking Perspective*, Princeton.
Connor, W. R. 1987. "Tribes, Festivals, and Processions: Civic, Ceremonial, and Political Manipulation in Archaic Greece," *JHS* 107, pp. 40–50.
Conomis, N. C. 1970. *Lycurgi: Oratio in Leocratem cum ceterarum Lycurgi orationum fragmentis*, Leipzig.
Conwell, D. H. 1992. "The Athenian Long Walls: Chronology, Topography, and Remains" [diss. University of Pennsylvania].
Conze, A. 1860. *Reise auf den Inseln der Thrakischen meeres*, Hanover.
Corbier, M. 1984. *Les mesures et l'histoire: Institute d'histoire moderne et contemporaine* [*CNRS: Cahiers de métrologie, Numéro speciale*], B. Garnier, ed., Paris.
Couilloud-Le Dinahet, M.-T. 1988. "Les magistrats grecs et l'approvisionnement des cités," in *Navires et commerces de la Mediterranée antique: Hommages à Jean Rougé* [*Cahiers d'histoire* 33], pp. 321–332.
Davies, J. K. 1971. *Athenian Propertied Families, 600–300 B.C.*, Oxford.
Davies, S. 1994. "Tithe-Collection in the Venetian Peloponnese, 1696–1705," *BSA* 89, pp. 443–455.
Day, J. 1942. *An Economic History of Athens under Roman Domination*, repr. New York, 1973.
Deonna, W. 1913. "Tables à mesures de capacités," *REA* 15, pp. 167–180.
———. 1938. *Le mobilier délien* [*Exploration Archéologique de Délos* XVIII], Paris.
De Ste. Croix, G. E. M. 1972. *The Origins of the Peloponnesian War*, London.
———. 1981. *The Class Struggle in the Ancient Greek World*, London.
Detienne, M., and J.-P. Vernant. 1989. *The Cuisine of Sacrifice among the Greeks*, P. Wissing, trans., Chicago.
Develin, R. 1989. *Athenian Officials, 684–321 B.C.*, Cambridge.
Dillery, J. 1996. "A List of Ptolemaic Cleruchs," in *P. Michigan Koenen (= P.Mich.XVIII), Michigan Texts Published in Honor of Ludwig Koenen*, C. E. Römer and T. Gagos, eds., Amsterdam, pp. 167–189.
Dover, K. J. 1981. "The Language of Classical Attic Documentary Inscriptions," *Transactions of the Philological Society: Oxford*, pp. 1–14. Reprinted in his *Greek and the Greeks*, Oxford 1987, pp. 31–41.
———. 1993. *Aristophanes* Frogs, Oxford.
———. 1997. *The Evolution of Greek Prose Style*, Oxford.

Dreher, M. 1995. *Hegemon und Symmachoi: Untersuchungen zum Zweiten Athenischen Seebund* [*Untersuchungen zur antiken Literatur und Geschichte* 46], Berlin.
Dübner, F. 1877. *Scholia Graeca in Aristophanem*, Paris.
Dunant, C., and J. Pouilloux. 1958. *Recherches sur l'histoire et les cultes de Thasos* II [*Études Thasiennes* V], Paris.
Edmonds, J. M. 1957. *The Fragments of Attic Comedy*, I, Leiden.
Engelmann, H., and D. Knibbe. 1989. "Das Zollgesetz der Provinz Asia: Eine neue Inschrift aus Ephesos," *EpAnt* 14.
Erxleben, H. 1974. "Die Rolle der Bevölkerungsklassen im Aussenhandel Athens im 4. Jahrhundert v. u. Z.," in *Hellenische Poleis*, I, E. C. Welskopf, ed., Darmstadt, pp. 460–520.
Fantasia, U. 1987. "Il grano di Leucone e le finanze di Atene: Nota a Demostene, 20, 33," *AnnPisa* 17, pp. 89–117.
Faraguna, M. 1992. *Atene nell'età di Alessandro: Problemi politici, economici, finanziari* [*MemLinc* ser. 9, no. 2.2], pp. 165–447.
Faraone, C. A. 1992. *Talismans and Trojan Horses*, New York.
Ferguson, W. S. 1938. "The Salaminioi of Heptaphylai and Sounion," *Hesperia* 7, pp. 1–74.
———. 1949. "Orgeonika," in *Hesperia* Supplement 8: *Commemorative Studies in Honor of Theodore Leslie Shear*, Princeton, pp. 130–163.
Feyel, M. 1942. *Contribution à l'épigraphie béotienne*, Le Puy.
Figueira, T. J. 1981. "Aeginetan Membership in the Peloponnesian League," *CP* 76, pp. 1–24.
———. 1986. "*Sitopolai* and *Sitophylakes* in Lysias' 'Against the Graindealers': Governmental Intervention in the Athenian Economy," *Phoenix* 40, pp. 149–171.
———. 1991. *Athens and Aigina in the Age of Imperial Colonization*, Baltimore.
———. 1993. *Excursions in Epichoric History: Aiginetan Essays*, Lanham, Md.
Finkelstein, M. I. 1935. "Ἔμπορος, Ναύκληρος, and Κάπηλος: A Prolegomena to the Study of Athenian Trade," *CP* 30, pp. 320–336.
Finley, M. I. 1985. *The Ancient Economy*, 2nd ed., Berkeley.
Fornara, C. W. 1971. *The Athenian Board of Generals from 501 to 404* [*Historia* Einzelschrift 16], Wiesbaden.
Fornara, C. W., and L. J. Samons II. 1991. *Athens from Cleisthenes to Pericles*, Berkeley.
Forsén, B., and G. R. Stanton. 1996. *The Pnyx in the History of Athens: Papers and Monographs of the Finnish Institute at Athens* II, Helsinki.
Fotiadis, M. 1995. "Modernity and the Past-Still-Present: Politics of Time in the Birth of Regional Archaeological Projects in Greece," *AJA* 99, pp. 59–78.
Foucart, P. 1884. "Note sur les comptes d'Éleusis sous l'archontat de Képhisophon, OL. 112, 4 (329/8)," *BCH* 8, pp. 194–216.
Foxhall, L., and H. Forbes. 1982. "Sitometria: The Role of Grain as a Staple Food in Classical Antiquity," *Chiron* 12, pp. 41–90.
Funke, P. 1980. *Homónoia und Arché: Athen und die griechische Staatenwelt vom Ende des peloponnesischen Krieges bis zum Königsfrieden* [*Historia* Einzelschrift 37], Wiesbaden.
Gabrielsen, V. 1990. "Trierarchic Symmories," *ClMed* 41, pp. 89–118.
———. 1994. *Financing the Athenian Fleet*, Baltimore.
Gagarin, M. 1981. *Drakon and Early Athenian Homicide Law*, New Haven.
Gallant, T. W. 1991. *Risk and Survival in Ancient Greece*, Stanford.
Gallo, L. 1983. "Alimentazione e classi sociali: Una nota su orzo e frumento in Grecia," *Opus* 2, pp. 449–472.
———. 1984. *Alimentazione e demografia della grecia antica*, Salerno.
Gargola, D. J. 1992. "Grain Distributions and the Revenue of the Temple of Hera on Samos," *Phoenix* 46, pp. 12–28.
Garlan, Y. 1974. *Recherches de poliorcétique grecque* [*BEFAR* 223], Athens.
Garnsey, P. D. A. 1985. "Grain for Athens," in *Crux: Essays in Greek History Presented to G. E. M. de Ste. Croix on His 75th Birthday*, P. A. Cartledge and F. D. Harvey, eds., London, pp. 62–75.
———. 1988. *Famine and Food Supply in the Graeco-Roman World*, Cambridge.
———. 1992. "Yield of the Land," in *Agriculture in Ancient Greece* [*Proceedings of the Seventh International Symposium at the Swedish Institute at Athens, 16–17 May, 1990*], B. Wells, ed., Stockholm, pp. 147–153.
Gauthier, P. 1976. *Un commentaire historique des* Poroi *de Xénophon* [*Centre de Recherches d'histoire et de philologie de la IV*e *Section de l'École pratique des Hautes Études* III, *Hautes Études du monde gréco-romaine* 8], Geneva.

———. 1981. "De Lysias à Aristote (Ath. pol., 51,4): Le commerce du grain à Athènes et les fonctions des sitophylaques," *RHDFE* 59, pp. 5–28.
———. 1987. "Nouvelles recoltes et grain nouveau: À propos d'une inscription de Gazaros," *BCH* 111, pp. 413–418.
———. 1989. *Nouvelles inscriptions de Sardes* II [*Centre de Recherches d'histoire et de philologie de la IVe Section de l'École pratique des Hautes Études* III, *Hautes Études du monde gréco-romaine* 15], Paris.
———. 1990. "L'inscription d'Iasos relative à l'*ekklesiastikon* (*I. Iasos* 20)," *BCH* 114, pp. 417–443.
———. 1993. "Sur l'institution du *misthos* de l'assemblée à Athènes (*Ath. Pol.* 41, 3)," in *Aristote et Athènes*, M. Piérart, ed., Paris, pp. 231–250.
Gera, G. 1975. *L'imposizione progressiva nell'antica Atene*, Rome.
Gernet, L. 1909. "L'approvisionnement d'Athènes en blé au Ve et au IVe siècle," in *Université de Paris: Bibliothèque de la Faculté des Lettres* 25, *Mélanges d'histoire ancienne*, Paris, pp. 271–391.
Gill, D. 1991. *Greek Cult Tables*, New York.
Gofas, D. C. 1969. "Les Carpologues de Thasos," *BCH* 93, pp. 337–370.
———. 1993. *Études d'histoire du droit grec des affaires: Antique, byzantin et post-byzantin* [*BAAH* 33], Athens.
Gomme, A. W. 1933a. *The Population of Athens in the Fifth and Fourth Centuries* B.C., Oxford.
———. 1933b. "A Forgotten Factor of Greek Naval Strategy," *JHS* 53, pp. 16–24.
Gomme, A. W., A. Andrewes, and K. J. Dover. 1945–1981. *A Historical Commentary on Thucydides*, 5 vols., Oxford.
Graham, A. J. 1983. *Colony and Mother City in Ancient Greece*, 2nd ed., Chicago.
Graindor, P. 1906. *Histoire de l'île de Skyros jusqu'en 1538*, Liège.
Graves, T. 1849. "On the Island of Skyros," *Journal of the Royal Society of London* 19, pp. 152–160.
Gruen, E. S. 1995. *The Last Generation of the Roman Republic*, 2nd ed., Berkeley.
Gschnitzer, F. 1958. *Abhängige Orte im griechischen Altertum* [*Zetemata* 17], Munich.
Guarducci, M. 1969. *Epigrafia greca*, II, Rome.
Hallof, K. 1990. "Der Verkauf konfiszierten Vermögens vor den Poleten in Athen," *Klio* 72, pp. 402–426.
Hammond, N. G. L. 1955. "Studies in Greek Chronology of the Sixth and Fifth Centuries B.C.," *Historia* 4, pp. 371–411.
———. 1983. "Army Transport in the Fifth and Fourth Centuries," *GRBS* 24, pp. 27–31.
Hansen, M. H. 1978a. "Demos, Ecclesia, and Dicasterion in Classical Athens," *GRBS* 19, pp. 127–146.
———. 1978b. "Nomos and Psephisma in Fourth-Century Athens," *GRBS* 19, pp. 315–330.
———. 1979a. "Did the Athenian *Ecclesia* Legislate after 403/2 B.C.?" *GRBS* 20, pp. 27–53.
———. 1979b. *Det Athenske demokrati i 4.Arh.f.kr. 5 Embedsmaendene*, Copenhagen.
———. 1983. *The Athenian Ecclesia: A Collection of Articles, 1976–1983*, Copenhagen.
———. 1985a. "Athenian *Nomothesia*," *GRBS* 26, pp. 341–371.
———. 1985b. *Demography and Democracy*, Herning.
———. 1986. "The Construction of Pnyx II and the Introduction of Assembly Pay," *ClMed* 37, pp. 89–98.
———. 1989. *The Athenian Ecclesia* II: *A Collection of Articles, 1983–1989*, Copenhagen.
———. 1990. "Diokles' Law (Dem. 24.42) and the Revision of the Athenian Corpus of Laws in the Archonship of Eukleides," *ClMed* 41, pp. 63–71.
———. 1991. *The Athenian Democracy in the Age of Demosthenes*, Oxford.
Hansen, M. V. 1984. "Athenian Maritime Trade in the 4th Century B.C.: Operation and Finance," *ClMed* 35, pp. 71–92.
Hanson, V. D. 1995. *The Other Greeks: The Family Farm and the Agrarian Roots of Western Civilization*, New York.
Harding, P. 1988. "Athenian Defensive Strategy in the Fourth Century," *Phoenix* 42, pp. 66–68.
Harris, E. M. 1989. "The Liability of Business Partners in Athenian Law: The Dispute between Lycon and Megacleides ([Dem.] 52.20–1)," *CQ* 39, pp. 339–343.
———. 1992. Review of D. M. MacDowell, ed., *Demosthenes: Against Meidias* (*Oration 21*), *CP* 87, pp. 71–80.
Harris, W. V. 1989. *Ancient Literacy*, Cambridge, Mass.
Harrison, A. R. W. 1971. *The Law of Athens: Procedure*, II, Oxford.
Hauben, H. 1975. "Agathokleia and Her Boats," *ZPE* 16, pp. 289–291.
Heichelheim, F. M. 1964. *An Ancient Economic History*, J. Stevens, trans., I–II, Leiden 1958–1964.
Henrichs, A. 1993. "The Tomb of Aias and the Prospect of Hero Cult in Sophokles," *ClAnt* 12, pp. 165–180.
Henry, A. S. 1996. "The Hortatory Intention in Athenian State Decrees," *ZPE* 112, pp. 105–119.

Hess, W. H. 1963. "Studies in the *Ekklesiazusae* of Aristophanes" [diss. Princeton University].
Hignett, C. 1952. *A History of the Athenian Constitution to the End of the Fifth Century* B.C., Oxford.
Hill, I. T. 1953. *The Ancient City of Athens*, London.
Hodkinson, S. 1988. "Animal Husbandry in the Greek Polis," in *Pastoral Economies in Classical Antiquity* [*PCPS* Supplement 14], Cambridge, pp. 35–74.
Hopper, R. J. 1979. *Trade and Industry in Classical Greece*, London.
Hornblower, S. 1991. *A Commentary on Thucydides: Vol. I, Books I–III*, Oxford.
How, W. W., and J. Wells. 1928. *A Commentary on Herodotus*, I–II, Oxford.
Hubbard, T. K. 1987. "Two Notes on the Myth of Aeacus in Pindar," *GRBS* 28, pp. 5–22.
Humphreys, S. 1985. "Lycurgus of Butadae: An Athenian Aristocrat," in *The Craft of the Ancient Historian: Essays in Honor of Chester G. Starr*, J. W. Eadie and J. Ober, eds., Lanham, Md., pp. 199–252.
Hunt, A. S. 1927. *Egypt Exploration Society: The Oxyrhynchus Papyri: Part XVII*, London.
Imhoof-Blumer, F. 1882. "Münzen der Kleruchen auf Imbros," *AM* 7, pp. 146–150.
Immerwahr, H. R. 1966. *Form and Thought in Herodotus*, Cleveland.
Isager, S., and M. H. Hansen. 1975. *Aspects of Athenian Society in the Fourth Century* B.C., Odense.
Isager, S., and J. E. Skydsgaard. 1992. *Ancient Greek Agriculture: An Introduction*, London.
Jardé, A. 1925. *Les céréales dans l'antiquité grecque* [*BEFAR* 130], Paris.
Jasny, N. 1944. *The Wheats of Classical Antiquity* [*The Johns Hopkins University Studies in Historical and Political Science* 62.3], Baltimore.
Jones, A. H. M. 1957. *Athenian Democracy*, Oxford.
———. 1974. "Taxation in Antiquity," in *The Roman Economy*, P. A. Brunt, ed., Oxford, pp. 151–185.
Judeich, W. *RE* I.1, 1893, col. 915 (Agyrrhios).
———. 1931. *Topographie von Athen*, 2nd ed., Munich.
Kalaitze, E. 1989. «Συμβολὴ εἰς τὴν ἱστορίαν Ἴμβρου καὶ Τενέδου», *Thrakika* 1, pp. 233–290.
Kallet-Marx, R. M., and R. S. Stroud. 1997. "Two Athenian Decrees Concerning Lemnos of the Late First Century B.C.," *Chiron* 27, pp. 155–194.
Kearns, E. 1989. *The Heroes of Attica* [*BICS* Supplement 57], London.
Knoepfler, D. 1981. "Argoura: Un toponyme eubéen dans la *Midienne* de Démosthène," *BCH* 105, pp. 289–329.
———. 1988. "L'intitulé oublié d'un compte des naopes béotiens," in *Comptes et inventaires dans la cité grecque* [*Actes du colloque international d'épigraphie tenu à Neuchâtel du 23 au 26 septembre 1986 en l'honneur de Jacques Tréheux*], D. Knoepfler, ed., Neuchâtel, pp. 263–294.
Köhler, U. 1891. "Die Halle der Athener in Delphi," *RhM* 46, pp. 1–8.
Körte, A. 1932. "Literarische Texte mit Ausschluss der christlichen," *AfPap* 10, pp. 217–237.
Koster, W. J. W. 1962. *Scholia in Aristophanem*, IV.3: *Jo. Tzetzae Commentarii in Aristophanem*, Groningen.
Koumanoudes, S. A. 1860. Ἐπιγραφαὶ Ἑλληνικαὶ κατὰ τὸ πλεῖστον Ἀνέκδοτοι, Athens.
———. 1862. Δύο λίθινα σκευὴ ἀγνώστου χρήσεως, Ἀρχ Ἐφ, cols. 23–26.
Kron, U. 1976. *Die zehn attischen Phylenheroen* [*AM* Beiheft 5], Berlin.
Kula, W. 1986. *Measures and Man*, R. Szreter, trans., Princeton.
Lambert, S. D. 1993. *The Phratries of Attica*, Ann Arbor.
———. 1997. *Rationes Centesimarum: Sales of Public Land in Lykourgan Athens*, Amsterdam.
Langdon, M. K. 1987. "An Athenian Decree Concerning Oropos," *Hesperia* 56, pp. 47–58.
———. 1994. "Public Auctions in Ancient Athens," in *Ritual, Finance, Politics: Athenian Democratic Accounts Presented to David Lewis*, R. Osborne and S. Hornblower, eds., Oxford, pp. 253–265.
Latte, K. 1953. *Hesychii Alexandrini Lexicon*, I, Copenhagen.
Lauffer, S., ed. 1989. *Griechenland: Lexikon der historischen Stätten*, Munich.
Lauter, H. 1982. "Zum Strassenbild in alt-Athen," *AntW* 13, pp. 44–52.
Lawton, C. L. 1995. *Attic Document Reliefs*, Oxford.
Leake, W. M. 1835. *Travels in Northern Greece*, London.
Lejeune, M. 1991. " 'Un huitième' dans le lexique métrologique grec," *REG* 104, pp. 198–201.
Lentz, A. 1867–1910. *Grammatici Graeci*, 6 vols., Leipzig.
Leppin, H. 1995. "Zur Entwicklung der Verwaltung öffentlicher Gelder im Athen des 4. Jahrhunderts v. Chr.," in *Die athenische Demokratie im 4. Jahrhundert v. Chr.*, W. Eder, ed., Stuttgart, pp. 557–571.
Lewis, D. M. 1955. "Notes on Attic Inscriptions (II)," *BSA* 50, pp. 1–36.

———. 1959. "Law on the Lesser Panathenaia," *Hesperia* 28, pp. 239–247.
———. 1960. "Apollo Delios," *BSA* 55, pp. 190–194.
Littman, R. J. 1988. "Greek Taxation," in *Civilization of the Ancient Mediterranean*, II, M. Grant and R. Kitzinger, eds., New York, pp. 795–808.
Lygouri-Tolia, E. 1985a. Δελτ 40 [1990] Β′, pp. 18–19.
———. 1985b. «Ανασκαφική έρευνα παρά την Πειραϊκή πύλη», *AAA* 18, pp. 137–142.
———. 1988. Δελτ 43 [1993] Β′, pp. 31–33.
Macan, R. W. 1895. *Herodotus: The Fourth, Fifth, and Sixth Books*, I, London.
MacDowell, D. M. 1962. *Andokides: On the Mysteries*, Oxford.
———. 1986. "The Law of Periandros about Symmories," *CQ* 36, pp. 438–449.
———. 1990. *Demosthenes: Against Meidias (Oration 21)*, Oxford.
McKechnie, P. 1989. *Outsiders in the Greek Cities in the Fourth Century B.C.*, London.
Maier, F. G. 1959–1961. *Griechische Mauerbauinschriften* [*Vestigia* 1–2], Heidelberg.
Malay, H. 1983. "A Royal Document from Aigai in Aiolis," *GRBS* 24, pp. 349–353.
Mark, I. S. 1993. *The Sanctuary of Athena Nike in Athens: Architectural Stages and Chronology* [*Hesperia* Supplement 26], Princeton.
Markle, M. M. 1985. "Jury Pay and Assembly Pay at Athens," in *Crux: Essays in Greek History Presented to G. E. M. de Ste. Croix on His 75th Birthday*, P. A. Cartledge and F. D. Harvey, eds., London, pp. 265–297.
Mattingly, H. B. 1964. "The Financial Decrees of Kallias IG I^2 91/92," *ProcAfrClSoc* 7, pp. 35–55.
Meritt, B. D. 1960. "Greek Inscriptions," *Hesperia* 29, pp. 1–77.
———. 1963. "Greek Inscriptions," *Hesperia* 32, pp. 1–56.
———. 1967. "Collectors of Athenian Tribute," *AJP* 88, pp. 29–32.
———. 1970. "The Election of Athenian Generals," *Klio* 52, pp. 277–282.
Merkelbach, R. 1991. "Weg mit Dir, Herakles, in die Feuershölle," *ZPE* 86, pp. 41–43.
Meyer, M. 1989. *Die griechischen Urkundenreliefs* [*AM* Beiheft 13], Berlin.
Michon, E., *DictAnt* IV.2, pp. 1176–1179 (*sekoma*).
Migeotte, L. 1984. *L'emprunt public dans les cités grecques: Recueil des documents et analyse critique*, Québec.
———. 1991. "Le pain quotidien dans les cités hellénistiques: À propos des fonds permanents pour l'approvisionnement en grain," *Cahiers du Centre G. Glotz* 2, Paris, pp. 19–41.
———. 1992. *Les souscriptions publiques dans les cités grecques* [*École Pratique des Hautes Études, IV*^{e} *Section, Sciences historiques et philologiques* III, *Hautes Études du monde gréco-romaine* 17], Geneva.
———. 1993. "Un fonds d'achat de grain à Coronée," in *Boeotia Antiqua* III: *Papers in Boiotian History, Institutions, and Epigraphy in Memory of Paul Roesch*, Amsterdam, pp. 11–23.
———. 1995. "Les finances publiques des cités grecques: Bilan et perspectives de recherche," *Topoi* 5, pp. 7–32.
———. 1996. "Les finances des cités grecques au-delà du primitivisme et du modernisme," in Ἐνεργεία: *Studies on Ancient History and Epigraphy Presented to H. W. Pleket*, J. H. M. Strubbe et al., eds., Amsterdam, pp. 79–96.
———. 1997. "Le contrôle des prix dans les cités grecques," in *Prix et formation des prix dans les économies antiques: Entretiens d'archéologie et d'histoire*, Saint-Betrand-de-Comminges.
———. Forthcoming, "Ventes." "Les ventes de grain public dans les cités grecques aux périodes classique et hellénistique," in *La mémoire perdue: Recherches sur l'administration romaine.*
———. Forthcoming, "Aspects." "Quelques aspects légaux et juridiques de l'affermage des taxes en grèce ancienne," in *Symposion 1997.*
Mikalson, J. 1975. *The Sacred and Civil Calendar of the Athenian Year*, Princeton.
Miller, Stephen G. 1978. *The Prytaneion: Its Function and Form*, Berkeley.
Millett, P. 1991. *Lending and Borrowing in Ancient Athens*, Cambridge.
Missiou, A. 1992. *The Subversive Oratory of Andokides*, Cambridge.
Montgomery, H. 1986. "'Merchants Fond of Corn': Citizens and Foreigners in the Athenian Grain Trade," *SymbOslo* 61, pp. 43–61.
Moritz, L. A. 1955. "Corn," *CQ* 49, pp. 135–141.
———. 1958. *Grain-Mills and Flour in Classical Antiquity*. Oxford.
Mossé, C. 1979. "Les symmories athéniennes," in *Points de vue sur la fiscalité antique*, H. van Effenterre, ed. [*Publications de la Sorbonne: Études* 14], Paris, pp. 31–42.

———. 1983. "The 'World of the *Emporium*' in the Private Speeches of Demosthenes," in *Trade in the Ancient Economy*, P. Garnsey, K. Hopkins, C. R. Whittaker, eds., Berkeley, pp. 53–63.
Moustoxydes, A. 1845. Ἀπόμνημα ἱστορικὸν περὶ τῆς νήσου Ἴμβρου, Istanbul.
Munn, M. 1993. *The Defense of Attica*, Berkeley.
Murray, O., and S. Price. 1990. *The Greek City from Homer to Alexander*, Oxford.
Musti, D. 1985. "Pubblico e privato nella democrazia periclea," *QuadUr* 2, pp. 7–17.
Naval Intelligence Division. 1945. *Geographical Handbook: Greece*, 3 vols., Andover.
Nenci, G. 1994. *Erodoto: Le Storie: Libro V*, Verona.
Nicolet, C. 1991. "Nouvelle observations sur la mosaïque des chevaux et son édifice à Carthage," *CRAI*, pp. 471–507.
———. 1994. *Le ravitaillement en blé de Rome*, Rome.
Noack, F. 1907. "Die Mauern Athens: Ausgrabungen und Untersuchungen," *AM* 32, pp. 123–160, 473–566.
Nolan, B. T. 1981. "Inscribing Costs at Athens in the Fourth Century B.C." [diss. Ohio State University].
Ober, J. 1985. *Fortress Attica* [*Mnemosyne* Supplement 84], Leiden.
———. 1989a. "Review Article: The Nature of Athenian Democracy," *CP* 84, pp. 322–334.
———. 1989b. *Mass and Elite in Democratic Athens*, Princeton.
Oehler, J. 1921. *RE* XI.1, col. 357 (Keryx).
Oikonomides, A. N. 1964. *The Two Agoras in Ancient Athens*, Chicago.
———. 1990. "The Aiakeion and Its Relation with the Tholos in the Later Agora," *AncW* 21, pp. 21–22.
Osborne, M. J. 1973. "The Stoichedon Style in Theory and Practice," *ZPE* 10, pp. 249–270.
Osborne, M. J., and S. G. Byrne. 1994. *A Lexicon of Greek Personal Names*, II, *Attica*, Oxford.
Osborne, R. 1988. "Social and Economic Implications of the Leasing of Land and Property in Classical and Hellenistic Greece," *Chiron* 18, pp. 279–323.
Pack, R. A. 1965. *Greek and Latin Literary Texts from Greco-Roman Egypt*, 2nd ed., Ann Arbor.
Parker, R. 1996. *Athenian Religion: A History*, Oxford.
Pélékidis, C. 1962. *Histoire de l'éphébie attique*, Paris.
Perrin, B. 1912. *Plutarch's Nicias and Alcibiades*, New York.
Philippaki, B. 1966. «Ἀνασκαφαὶ ἐντὸς τῆς περιμετρικῆς ζώνης τῶν Ἀθηνῶν», Δελτ 21 [1968] Β′, pp. 55–57.
Philippson, A. 1901. *Beiträge zur Kenntnis der griechischen Inselwelt*, Gotha.
Pickard-Cambridge, A. W. 1968. *The Dramatic Festivals of Athens*, 2nd ed., rev. J. Gould and D. M. Lewis, Oxford.
Piérart, M. 1971. "Les ΕΥΘΥΝΟΙ athéniens," *AntCl* 40, pp. 526–573.
———. 1974. "À propos de l'élection des stratèges athéniens," *BCH* 98, pp. 125–146.
Pleket, H. W. 1964. *Epigraphica I: Texts on the Economic History of the Greek World* [*Textus Minores* XXXI], Leiden.
Podlecki, A. J. 1977. "Herodotus in Athens?" in *Greece and the Eastern Mediterranean in Ancient History and Prehistory: Studies Presented to Fritz Schachermeyr*, K. Kinzl, ed., Berlin, pp. 246–265.
Pouilloux, J. 1954. *Recherches sur l'histoire et les cultes de Thasos* [*Études Thasiennes* III], Paris.
Prinz, F. 1979. *Gründungsmythen und Sagenchronologie* [*Zetemata* 72], Munich.
Pritchett, W. K. 1940. "The Term of Office of the Athenian Strategoi," *AJP* 61, pp. 469–474.
———. 1953. "The Attic Stelai: Part I," *Hesperia* 22, pp. 225–299.
———. 1954. "Sales Taxes in Ancient Athens," *Archaeology* 7, pp. 112–113.
———. 1956a. "The Attic Stelai: Part II," *Hesperia* 25, pp. 178–317.
———. 1956b. "Fourth-Century Athenian Sales Taxes," *CP* 51, pp. 100–102.
———. 1974. *The Greek State at War*, II, Berkeley.
———. 1980. *Studies in Ancient Greek Topography: Part III (Roads)* [*University of California Publications: Classical Studies* 22], Berkeley.
———. 1991. *The Greek State at War*, V, Berkeley.
———. 1992. *Studies in Ancient Greek Topography: Part VIII*, Amsterdam.
———. 1993. *The Liar School of Herodotus*, Amsterdam.
———. 1996. *Greek Archives, Cults, and Topography*, Amsterdam.
Race, W. H. 1987. "Pindaric Encomium and Isocrates' *Evagoras*," *TAPA* 117, pp. 131–155.
Raepset, G. 1988. "Charrettes en terre cuite de l'époque archaïque de Corinthe," *AntCl* 57, pp. 56–88.
Rankin, D. I. 1988. "The Mining Lobby at Athens," *Ancient Society* 19, pp. 189–205.

Rathbone, D. W. 1983. "The Weight and Measurement of Egyptian Grains," *ZPE* 53, pp. 265–275.
Raubitschek, A. E. 1941. "The Heroes of Phyle," *Hesperia* 10, pp. 284–295.
———. 1949. *Dedications from the Athenian Akropolis*, Cambridge, Mass.
Rauh, N. K. 1993. *The Sacred Bonds of Commerce*, Amsterdam.
Reger, G. 1993. "The Public Purchase of Grain on Independent Delos," *ClAnt* 12, pp. 299–334.
———. 1994. *Regionalism and Change in the Economy of Independent Delos, 314–167 B.C.*, Berkeley.
Reinmuth, O. 1955. "The Ephebic Inscription, Athenian Agora I 286," *Hesperia* 24, pp. 220–239.
Rennie, W. 1909. *Aristophanes* Acharnians, London.
Rhodes, P. J. 1972. *The Athenian Boule*, Oxford.
———. 1980. "Athenian Democracy after 403 B.C.," *CJ* 75, pp. 305–323.
———. 1981. *A Commentary on the Aristotelian* Athenaion Politeia, Oxford.
———. 1982. "Problems in Athenian *Eisphora* and Liturgies," *AJAH* 7 [1985], pp. 1–19.
———. 1987. "Nomothesia in Classical Athens," in *L'Educazione Giuridica*, V, *Modelli di legislatore e scienza della legislazione*, II; *Modelli storici e comparativi*, A. Giuliani and N. Picardi, eds., Perugia, pp. 5–26.
Riccobono, S., et al., eds. 1941. *Fontes iuris romani Antejustiniani*, I, Florence.
Rickman, G. 1980. *The Corn Supply of Ancient Rome*, Oxford.
Rosivach, V. 1992. "The *rationes centesimarum*: (IG 2^2 1594–1603+)," *Eirene* 28, pp. 49–61.
Rothwell, K. S., Jr. 1990. *Politics and Persuasion in Aristophanes'* Ecclesiazusae [*Mnemosyne* Supplement 111], Leiden.
Ruschenbusch, E. 1978. "Die Athenischen Symmorien des 4. Jh.v.Chr.," *ZPE* 31, pp. 275–284.
———. 1979. "Die Einführung des Theorikon," *ZPE* 36, pp. 303–308.
———. 1987. "Symmorienprobleme," *ZPE* 69, pp. 75–81.
———. 1990. "Die Zahl der Athenischen Trierarchien in der Zeit nach 340 v. Chr.," *ClMed* 41, pp. 79–88.
Rusten, J. 1983. "ΓΕΙΤΩΝ ΗΡΩΣ: Pindar's Prayer to Heracles (*N.* 7.86–101) and Greek Popular Religion," *HSCP* 87, pp. 289–297.
Rutherford, I. 1992. "Two Heroic *Prosodia:* A Study of Pindar, 'Paeans XIV–XV,' " *ZPE* 92, pp. 59–72.
Sallares, R. 1991. *The Ecology of the Ancient Greek World*, Ithaca.
Salomon, N. 1997. *Le cleruchie di Atene: Caratteri e funzione* [*Studi e testi di storia antica diretti da Mauro Moggi* 6], Florence.
Sartori, F. 1983. "Aristofane e Agirrio nel 405 a.C.," in *Althistorische Studien Hermann Bengtson zum 70. Geburtstag dargebracht von Kollegen und Schülern*, H. Heinen et al., eds. [*Historia* Einzelschrift 40], Wiesbaden, pp. 56–77.
Schlaifer, R. 1940. "Notes on Athenian Private Cults," *HSCP* 51, pp. 233–260.
Schmidt, M. 1858. *Hesychii Alexandrini Lexicon*, I, Jena.
Schmitt-Pantel, P. 1992. *La cité au banquet: Histoire des repas publics dans les cités grecques* [*Collection École française de Rome* 157], Rome.
Schmitz, W. 1988. *Wirtschaftliche Prosperität, soziale Integration und die Seebundpolitik Athens* [*Quellen und Forschungen zur Antiken Welt* 1], Munich.
Schwahn, W. 1934. *RE* VA, col. 245 (Tele).
———. 1936a. *RE* XXXIII, cols. 843–845 (Nomos telonikos).
———. 1936b. *RE* VIA, col. 575 (Thrasybulos no. 5).
Schwartz, J. 1969. "Hérodote et Périclès," *Historia* 18, pp. 367–370.
Schweigert, E. 1940. "Greek Inscriptions," *Hesperia* 9, pp. 309–357.
Seager, R. 1967. "Thrasybulus, Conon, and Athenian Imperialism, 396–386 B.C.," *JHS* 87, pp. 95–115.
Sealey, R. 1956. "Callistratos of Aphidna and His Contemporaries," *Historia* 5, pp. 178–203.
———. 1987. *The Athenian Republic*, University Park, Penn.
———. 1993. *Demosthenes and His Time*, Oxford.
Shear, T. L., Jr. 1970. "The Monument of the Eponymous Heroes in the Athenian Agora," *Hesperia* 39, pp. 145–222.
———. 1978. *Kallias of Sphettos and the Revolt of Athens in 286 B.C.* [*Hesperia* Supplement 17], Princeton.
———. 1993. "The Persian Destruction of Athens: Evidence from Agora Deposits," *Hesperia* 62, pp. 383–482.
Shrimpton, G. 1987. "Theopompos on the Death of Evagoras I," *AHB* 1, pp. 105–111.
Sickinger, J. P. 1994a. "Inscriptions and Archives in Classical Athens," *Historia* 43, pp. 286–296.

———. 1994b. Review of R. Thomas, *Literacy and Orality in Ancient Greece*, *CP* 89, pp. 273–278.
Simon, D. 1964. *Studien zur Praxis der Stipulation Klausel* [*MünchBeitr* 48], Munich.
Sinclair, R. K. 1988. *Democracy and Participation in Athens*, Cambridge.
Smarczyk, B. 1990. *Untersuchungen zur Religionspolitik und politischen Propaganda Athens im Delisch-Attischen Seebund* [*Quellen und Forschungen zur Antiken Welt* 5].
Spathari, E. 1982. Δελτ 37 [1989] Β′, pp. 23–24.
Strauss, B. S. 1987. *Athens after the Peloponnesian War: Class, Faction, and Policy, 403–386 B.C.*, Ithaca.
Stroud, R. S. 1971. "Inscriptions from the North Slope of the Acropolis, I," *Hesperia* 40, pp. 146–204.
———. 1974. "An Athenian Law on Silver Coinage," *Hesperia* 43, pp. 157–188.
———. 1994. "The Aiakeion and Tholos of Athens in POxy 2087," *ZPE* 103, pp. 1–9.
Strubbe, J. H. M. 1989. "The Sitonia in the Cities of Asia Minor under the Principate," *EpAnt* 13, pp. 99–122.
Tarbell, F. B. 1891. "A 'mensa ponderaria' from Assos," *AJA* 7, pp. 440–443.
Taylor, M. C. 1995. "A Fourth Century Honorary Decree of the Athenian Demos on Salamis," *ZPE* 107, pp. 289–295.
———. 1997. *Salamis and the Salaminioi*, Amsterdam.
Thomas, R. 1989. *Oral Tradition and Written Record in Classical Athens*, Cambridge.
———. 1992. *Literacy and Orality in Ancient Greece*, Cambridge.
Thompson, D. J. 1983. "Nile Grain Transport under the Ptolemies," in *Trade in the Ancient Economy*, P. Garnsey and K. Hopkins, eds., Berkeley, pp. 64–75.
Thompson, H. A. 1948. "The Excavations of the Athenian Agora, Twelfth Season: 1947," *Hesperia* 17, pp. 149–196.
———. 1953. "Excavations in the Athenian Agora: 1952," *Hesperia* 22, pp. 25–56.
———. 1954. "Excavations in the Athenian Agora: 1953," *Hesperia* 23, pp. 31–67.
———. 1955. "Activities in the Athenian Agora: 1954," *Hesperia* 24, pp. 50–71.
———. 1988. "Building for a More Democratic Society: The Athenian Agora after Ephialtes," in Πρακτικά του XII Διεθνούς Συνεδρίου Κλασικής Αρχαιολογίας 4, Athens, pp. 198–203.
Thompson, H. A., and R. L. Scranton. 1943. "Stoas and City Walls on the Pnyx," *Hesperia* 12, pp. 269–383.
Thompson, W. E. 1970. "Notes on Attic Demes," *Hesperia* 39, pp. 64–67.
———. 1978. "The Athenian Investor," *RSC* 26, pp. 403–423.
———. 1982. "The Athenian Entrepreneur," *AntCl* 51, pp. 53–85.
Thomsen, R. 1964. *Eisphora*, Copenhagen.
Threatte, L. 1980. *The Grammar of Attic Inscriptions*, I: *Phonology*, Berlin.
———. 1996. *The Grammar of Attic Inscriptions*, II: *Morphology*, Berlin.
Threpsiades, I. 1953. «Ἀνασκαφικαὶ ἔρευναι ἐν Ἀθήναις», Πρακτικά, pp. 61–71.
Threpsiades, I., and E. Vanderpool. 1964. "Themistokles' Sanctuary of Artemis Aristoboule," Δελτ 19 [1965] Α′, pp. 26–36.
Todd, S. 1990. "Lady Chatterley's Lover and the Attic Orators," *JHS* 110, pp. 146–173.
Tournefort, J. de. 1741. *A Voyage into the Levant*, London.
Tracy, S. V. 1975. *The Lettering of an Athenian Mason* [*Hesperia* Supplement XV], Princeton.
———. 1984. "Greek Inscriptions from the Athenian Agora," *Hesperia* 53, pp. 369–377.
———. 1990. *Attic Letter-Cutters of 229 to 86 B.C.*, Berkeley.
———. 1995. *Athenian Democracy in Transition: Attic Letter-Cutters of 340 to 290 B.C.*, Berkeley.
Traill, J. S. 1975. *The Political Organization of Attica* [*Hesperia* Supplement 14], Princeton.
———. 1986. *Demos and Trittys*, Toronto.
———. 1994. *Persons of Ancient Athens*, I, *A to Alexandros*, Toronto.
Travlos, J. 1960. Πολεοδομικὴ Ἐξέλιξις τῶν Ἀθηνῶν, Athens.
———. 1971. *Pictorial Dictionary of Ancient Athens*, London.
———. 1988. *Bildlexikon zur Topographie des Antiken Attika*, Tübingen.
Tréheux, J. 1987. "ΚΟΙΝΟΝ," *REA* 89, pp. 39–46.
———. 1991. "Retour à Imbros et à Samothrace," *Hellénika Symmikta: Études d'archéologie classique* VII, Nancy, pp. 143–149.
Trevett, J. C. 1990. "P. Oxy. 2537 and Isokrates' Trapeziticus," *ZPE* 81, pp. 22–26.
Tsountas, C. 1883. Λογοδοσία ἀρχῶν τοῦ ἐν Ἐλευσῖνι ἱεροῦ, Ἀρχ Ἐφ, cols. 109–131, 253–264.

Tuplin, C. 1993. *The Failings of Empire* [*Historia* Einzelschrift 76], Stuttgart.

Turner, L. A. 1994. "The History, Monuments, and Topography of Ancient Lebadeia in Boeotia, Greece" [diss. University of Pennsylvania].

Uguzzoni, A., and F. Ghinatti. 1968. *Le tavole greche di eraclea: Università degli Studi di Padova* [*Pubblicazioni dell'Istituto di Storia Antica* 7], Rome.

Ussher, R. G. 1973. *Aristophanes:* Ecclesiazusae, Oxford.

van Berchem, D. 1991. "Commerce et écriture," *MusHelv* 48, pp. 129–145.

van Effenterre, H. 1960. "Un *sékoma* crétois," *BCH* 84, pp. 233–241.

van Leeuwen, J. 1940. *Aristophanes,* Equites, Leiden.

Velissaropoulos, J. 1977. "Le monde de l'emporion," *Dialogues de l'histoire ancienne* 3, pp. 61–85.

———. 1980. *Les nauclères grecs* [*Centre de Recherches d'histoire et de philologie de la IVe Section de l'École pratique des Hautes Études* III, *Hautes Études du monde gréco-romaine* 9], Geneva.

Veyne, P. 1985. "Une inscription Dionysiaque peu commune," *BCH* 109, pp. 621–625.

Wachsmuth, C. 1890. *Die Stadt Athen in Altertum*, II, Leipzig.

Walbank, F. W. 1957–1979. *A Historical Commentary on Polybius*, 3 vols., Oxford.

Walbank, M. B. 1982a. "The Confiscation and Sale by the Poletai in 402/1 B.C. of the Property of the Thirty Tyrants," *Hesperia* 51, pp. 74–98.

———. 1982b. "An Athenian Decree Re-considered: Honours for Aristoxenos and Another Boiotian," *EMC* 26, pp. 259–274.

———. 1983. "Leases of Sacred Properties in Attica, Part I," *Hesperia* 52, pp. 100–135; Part II, pp. 177–199; Part III, pp. 200–206; Part IV, pp. 207–231.

———. 1987. "Athens Honors Prono(u)s Son of Thyion of Samos," *ZPE* 69, pp. 265–268.

———. 1994. "A *Lex Sacra* of the State and of the Deme of Kollytos," *Hesperia* 63, pp. 233–239.

Wallace, R. W. 1992. "Charmides, Agariste, and Damon: Andokides 1.16," *CQ* 42, pp. 328–335.

Wallinga, H. T. 1964. "Nautika (I): The Unit of Capacity for Ancient Ships," *Mnemosyne* 17, pp. 1–40.

———. 1993. *Ships and Sea-Power before the Great Persian War* [*Mnemosyne* Supplement 121], Leiden.

Walter, H. 1993. *Aigina: Die archäologische Geschichte einer griechischen Insel*, Munich.

Walter-Karydi, E. 1987. *Alt-Ägina* II, ii: *Die Äginetische Bildhauerschule*, Mainz.

———. 1994. "Das Theorion von Aigina: Zum Apollonkult auf Aigina," *AA*, pp. 125–138.

Welles, C. B. 1934. *Royal Correspondence in the Hellenistic Period*, New Haven.

———. 1938. "New Texts from the Chancery of Philip V," *AJA* 42, pp. 245–260.

Welter, G. 1938. *Aigina*, Berlin.

———. 1962. Αἴγινα, Athens.

West, W. C. 1989. "The Public Archives in Fourth-Century Athens," *GRBS* 30, pp. 529–543.

Whitehead, D. 1982. "Notes on Athenian Demarchs," *ZPE* 47, pp. 37–42.

———. 1986. *The Demes of Attica, 508/7–ca. 250 B.C.: A Political and Social Study*, Princeton.

Wilamowitz-Moellendorff, U. v. 1893. *Aristoteles und Athen*, I–II, Berlin.

Wilhelm, A. 1915. *Neue Beiträge zur griechischen Inschriftenkunde* IV [*SBWien*, Bd. 179.6], Vienna.

———. 1923. "Zu Inschriften aus Kleinasien," in *Anatolian Studies Presented to Sir William Mitchell Ramsay*, W. H. Buckler and W. M. Calder, eds., Manchester, pp. 415–439.

Williams, D. 1987. "Aegina, Aphaia-Tempel XI: The Pottery from the Second Limestone Temple and the Later History of the Sanctuary," *AA*, pp. 629–680.

Wilson, C. H. 1970. "Athenian Military Finances, 378/7 to the Peace of 375," *Athenaeum* 48, pp. 302–326.

Wolff, H. J. 1941. "The *Praxis*-Provision in Papyri-Contracts," *TAPA* 72, pp. 418–438.

Woodford, S. 1971. "Cults of Heracles in Attica," in *Studies Presented to George M. A. Hanfmann*, D. G. Mitten et al., eds., Cambridge, Mass., pp. 211–227.

Wurster, W. W. 1974. *Alt Ägina* I, i: *Der Apollontempel*, Mainz.

Wycherley, R. E. 1959. "Two Athenian Shrines," *AJA* 63, pp. 67–72.

———. 1978. *The Stones of Athens*, Princeton.

Young, R. S. 1951. "An Industrial District of Ancient Athens," *Hesperia* 20, pp. 135–288.

Youtie, H. C. 1967. "Publicans and Sinners," *ZPE* 1, pp. 1–20.

Zschietzschmann, W. 1973. *RE*, Suppl. XIII, col. 72 (Athenai).

Zunker, A. 1988. *Untersuchungen zur Aiakidensage auf Aigina* [*Dissertationen Philosophische Reihe* 4], St. Ottilien.

INDEXES

LIST OF PASSAGES CITED

INSCRIPTIONS CITED

IMPORTANT GREEK WORDS

GENERAL INDEX

"See under" references to Greek terms may be found in the *Index of Important Greek Words.*